"十四五"职业教育部委级规划教材

教育部国家职业教育专业教学资源民族文化传承与创新子库"中国丝绸技艺民族文化传承与创新"配套双语教材

江苏省高等职业院校高水平专业群"纺织品检验与贸易"配套教材

缂丝织造技艺
Kossu Fabrics

黄紫娟◎主　编

詹停停　范玉明　蔡霞明◎副主编

刘　丽◎译

U0161645

中国纺织出版社有限公司

内 容 提 要

本书详细介绍了缂丝织造技艺的历史渊源、发展历程、缂丝织造技艺的工艺过程、主要技法、不同类型的缂丝产品、缂丝传承与创新以及非物质文化遗产代表性传承人。

本书可作为职业院校艺术类、纺织服装类及相关专业的教材，也可供对缂丝感兴趣的工艺美术工作者和社会读者参考，同时也可作为"中文＋技能"培训书籍。

This textbook mainly introduces the historical origin and development of kossu weaving, the entire process of kossu weaving, the main techniques, different types of kossu products, kossu inheritance and innovation, and the representative inheritors of this intangible cultural heritage. This textbook adopts a bilingual format of Chinese and English translation, so that more people can understand and know more about the ancient craft of kossu, the intangible cultural heritage.

This textbook is available for vocational students of art,textile and apparel majors of higher vocational institutions, and it can also be useful as a reference for arts and crafts workers, other readers who are interested in kossu. Meanwhile,it can also be used as a training book for chinese and skills.

图书在版编目（CIP）数据

缂丝织造技艺 =Kossu Fabrics：汉文、英文 / 黄紫娟主编；詹停停，范玉明，蔡霞明副主编；刘丽译. -- 北京：中国纺织出版社有限公司，2022.12
"十四五"职业教育部委级规划教材　教育部国家职业教育专业教学资源民族文化传承与创新子库"中国丝绸技艺民族文化传承与创新"配套双语教材　江苏省高等职业院校高水平专业群"纺织品检验与贸易"配套教材
ISBN 978-7-5180-9515-5

Ⅰ. ①缂… Ⅱ. ①黄… ②詹… ③范… ④蔡… ⑤刘… Ⅲ. ①缂丝—刺绣—丝织工艺—高等职业教育—教材—汉、英 Ⅳ. ①TS145.3

中国版本图书馆 CIP 数据核字（2022）第 068875 号

责任编辑：孔会云　沈　靖　　责任校对：寇晨晨　　责任印制：王艳丽

中国纺织出版社有限公司出版发行
地址：北京市朝阳区百子湾东里A407号楼　邮政编码：100124
销售电话：010—67004422　传真：010—87155801
http://www.c-textilep.com
中国纺织出版社天猫旗舰店
官方微博 http://weibo.com/2119887771
北京通天印刷有限责任公司印刷　各地新华书店经销
2022年12月第1版第1次印刷
开本：787×1092　1/16　印张：13.25
字数：262千字　定价：88.00元

前 言 / **Foreword**

　　缂丝织造技艺源于缂毛技艺，它真实地反映了历史变迁中不同时期的世俗与风尚，由此树立起文明的标尺，同时体现人文、启人智慧，充实和提高精神生活。缂丝艺人将手艺的精湛技巧与艺术的丰富想象完美结合，使技进乎于道，使艺净化人生。缂丝织造技艺经过不同历史时期的更替与沉积，形成了丰富的艺术风格和文化特征，已经成为具有极高艺术审美价值的瑰宝。

　　2006年5月20日，苏州缂丝织造技艺经国务院批准列入第一批国家级非物质文化遗产名录；2009年9月30日，缂丝织造技艺作为中国蚕桑丝织技艺的一部分入选世界非物质文化遗产名录；2016年1月14日，南通缂丝织造技艺被列入第四批江苏省非物质文化遗产名录；2021年6月10日，定州缂丝织造技艺被列入第五批国家级非物质文化遗产名录。2021年，中共中央办公厅、国务院办公厅印发《关于进一步加强非物质文化遗产保护工作的意见》指出，非物质文化遗产是中华优秀传统文化的重要组成部分，是中华文明绵延传承的生动见证，是连结民族情感、维系国家统一的重要基础。保护好、传承好、利用好非物质文化遗产，对于延续历史文脉、坚定文化自信、推动文明交流互鉴、建设社会主义文化强国具有重要意义。

　　在对缂丝历史发展进行较为系统的梳理后，根据缂丝织造技艺目前发展的状况及特点，全书设置了五个章节，从缂丝的起源与发展到具体的工艺流程、织机与辅助工具、缂丝产品、缂丝的传承与创新、缂丝非遗大师介绍，对缂丝织造技艺进行了系统、全面的介绍。

　　本书由苏州经贸职业技术学院黄紫娟主编并统稿。编写人员及分工如下：第一章由黄紫娟、詹停停编写；第二章由黄紫娟、范玉明编写；第三章由黄紫娟编写；第四章由黄紫娟、蔡霞明编写；第五章由黄紫娟、曹美姐编写；全书英文由西安工程大学刘丽翻译，胡伟华总译审。

　　本书受到教育部国家职业教育专业教学资源民族文化传承与创新子库"中国丝绸技艺民族文化传承与创新"和江苏省高等职业教育高水平专业群"纺织品检验与贸易"资助，同时得到了中国纺织出版社有限公司、苏州经贸职业技术学院纺织服装与艺术传媒学院、苏州工业园区仁和织绣有限公司、吴文康、祯彩堂陈文、王金山大师工作室的大力支持，在此表示衷心感谢。

　　本书在编写过程中参考了大量缂丝织造技艺相关的书籍、资料和图文，在此对这些书籍、资料和图文的作者致以诚挚的感谢。由于作者水平有限，书中难免存在疏漏和不足之处，欢迎广大读者、专家批评指正。

编者

2022年3月

翻译前言 / Translation Preface

《缂丝织造技艺》可为对缂丝织造感兴趣的读者以及缂丝织造行业的从业者提供参考。本书采用中英文互译的形式，有助于缂丝织造技艺被国内外更多的人了解并熟知。

本教材的英文由西安工程大学人文社会科学学院刘丽副教授负责组织翻译，协助人员有丁远（第一章第一、第二节）、曹雨菲（第一章第三、第五节）、徐倩（第一章第六、第七节）、贾亚菲（第二章、第五章）、肖芳英（第三章、第五章）和王梦雪（第四章、第五章）。此外，闵书嫣、丁远、闫佳颖参与了本书的校对。西安工程大学人文社会科学学院胡伟华教授帮助审查了所有内容的翻译，并提出了宝贵的修改建议。

在此诚挚感谢所有翻译、校对和审稿人的努力和辛勤工作！

Kossu Fabrics can serve as a reference for the readers who are interested in kossu fabrics as well as practitioners in the kossu textile industry. With comprehensive content and complete structure and fluent language, the textbook pays much attention to integrating theoretical knowledge with practice.

The translation of this textbook is organized and conducted by Li Liu, associated professor of School of Humanities and Social Sciences, Xi'an Polytechnic University, with the assistance of Yuan Ding (Sections 1 and 2 of Chapters 1), Yufei Cao (Sections 3 and 5 of Chapters 1), Qian Xu (Sections 6 and 7 of Chapters 1), Yafei Jia (Chapters 2 and 5), Fangying Xiao (Chapters 3 and 5) and Mengxue Wang (Chapters 4 and 5). Shuyan Min, Yuan Ding and Jiaying Yan participated in the proofreading of the text book. Dr. Weihua Hu, professor of School of Humanities and Social Sciences of Xi'an Polytechnic University, helped to review all the translations and put forward valuable suggestions for revision.

We are grateful for the efforts and hard work of all the translators, proofreaders and reviewers!

译者
2022年8月

目　录／**Contents**

◎ 第一章
缂丝的历史渊源
Historical Origin of Kossu Fabrics

◎ 概述/Introduction

缂丝是世界非物质文化遗产，因其工艺复杂、生产量少，导致接触并了解的人群十分有限。为了让更多的人深入地了解缂丝，传承并发扬缂丝非遗文化，本章将阐述缂丝的起源与传播，详细介绍唐、宋、辽、西夏、元、明、清、近代等不同时期缂丝的发展，分别从产生的历史背景、产品的特点、出现的名家名作等方面进行介绍。本章根据任务要求，按时间顺序介绍缂丝的历史渊源，并从材质、工艺、花纹等方面进行对比，使读者对缂丝的发展有清晰的认识。

Kossu is a world intangible cultural heritage. Its complicated technology and low production limit the population who can contact and understand it. In order to make learners have a deeper understanding of what is kossu, inherit and carry forward it, this chapter will start from the origin and emergence of kossu, and then describe the development of kossu in different dynasties including the Tang, Song, Liao, Kingdom of Xia, Yuan, Ming, Qing and modern times. The historical background, characteristics of product, famous craftsmen and works are described respectively. According to the task requirements, this chapter introduces the historical origin of kossu, and compares them from the details of material, technology and pattern in chronological order. So learners can have a clear understanding of the development of kossu.

◎ 思维导图/Mind Map

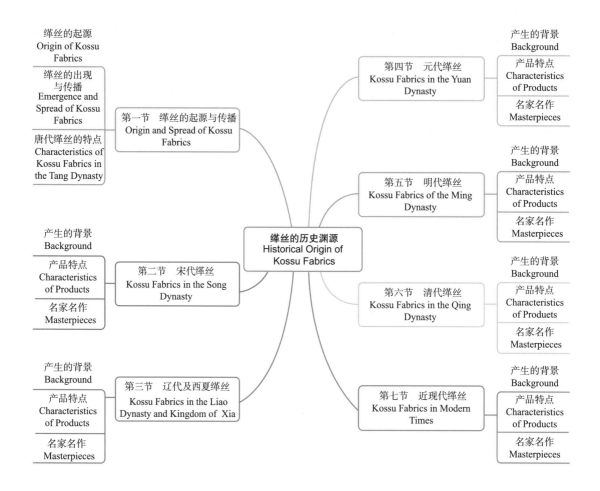

◎ 教学目标/Teaching Objectives

知识目标/Knowledge Goals

1. 了解缂丝的概念/The Concept of Kossu Fabrics

2. 掌握缂丝出现的历史条件/Historical Conditions for the Emergence of Kossu Fabrics

3. 掌握唐代缂丝出现的条件及其特点/Emergence Conditions and Characteristics of Kossu Fabrics in the Tang Dynasty

4. 掌握宋、元、明、清不同时期缂丝的发展脉络、特点和名家名作/Development, Characteristics and Masterpieces of Kossu Fabrics in different dynasties including the Song, Yuan, Ming and Qing Dynasties

5. 了解近现代缂丝发展/Modern Development of Kossu Fabrics

技能目标/Skill Goals

1. 具备分析缂丝出现的历史背景的能力/Analyzing the Historical Background of the Emergence of Kossu Fabrics

2. 能分析各个朝代缂丝的特点/Analyzing the Characteristics of Kossu Fabrics in Different Dynasties

素质目标/Quality Goals

1. 具备良好的审美能力/Possessing Good Aesthetic Ability
2. 树立创新意识和创新精神/Developing Innovative Consciousness and Spirit

思政目标/Ideological and Political Goals

1. 树立实事求是的历史唯物主义观/Establishing a Historical Materialism View of Seeking Truth from Facts
2. 弘扬中华优秀传统文化/Promoting the Excellent Traditional Chinese Culture

第一节　缂丝的起源与传播/Origin and Spread of Kossu Fabrics

一、缂丝的起源/Origin of Kossu Fabrics

缂丝是我国古老的丝织工艺，以平纹组织为基础、"通经断纬"织造，生桑蚕丝作经、彩色熟桑蚕丝作纬，利用多色纬线的不断变化，用一种竹舟形的小梭，按图案局部挖织形成花纹的织物。"通经断纬"即纬线并不贯穿整幅织物，而是只在花纹局部内回纬织制，因而花纹与素地之间、色与色之间的交界处呈现一些互不相连的断痕，如刀刻一般，"承空视之，如雕镂之象"，画面文雅古朴，极具赏玩、收藏价值，因此，缂丝被称为"织中之圣"。

Kossu is an ancient silk weaving technology in China. Based on plain weave, it is woven by the method "warp-passing and weft-breaking", with raw mulberry silk as the warp and colored processed mulberry silk as the weft. With the constant change of multicolored weft threads, a small boat-shaped bamboo shuttle is used partially to dig and weave patterns according to the model pattern. "Warp-passing and weft-breaking" means that the weft thread does not run through the whole fabric, but is only woven back in the pattern parts. Therefore, like a knife carving, there are some disconnected breaks at the junction between the patterns and the plain parts, as well as between different colors. "Seeing from above, it is like a carved image." Being elegant and simple, the kossu fabric is honored as "the Holy of fabrics", having great value for enjoyment and collection.

　　缂丝技艺源于以毛线为纬的缂毛织物。缂毛最早起源于今西亚地区，随着纺织技术的发展，缂毛技艺传播到欧洲等地区，并从西向东传播到中国新疆以及内陆地区。我国境内最早的缂毛织物是19世纪初期由英国探险家斯坦因在楼兰发现，中国的考古工作者先后在巴楚县脱库孜沙来遗址、洛浦县山普拉、且末县札滚鲁克等地发现缂毛织物，地理上位于新疆塔里木盆地周围的绿洲地带，如图1-1-1和图1-1-2所示。

The technique of kossu originated from the weaving of woolen fabrics with wool as the weft. The wool of kossu fabrics was first found in present western Asia. With the development of textile technology, wool weaving skills spread to Europe.The skills then spread from west to east to Xinjiang and inland areas of China. The earliest woolen fabrics found in China were discovered in Loulan by British explorer Stein in the early 19th century. Chinese archaeologists have successively found woolen fabrics in Tou Kuzishala, Bachu County, Shanpula, Luopu County and Zhagunluk, Qiemo County, which are geographically located in the oasis zone around Tarim Basin in Xinjiang, as shown in Figures 1-1-1 and 1-1-2.

缂毛

图1-1-1　洛浦县山普拉缂毛武士像
Warrior's Woolen Image in Sampula,
Luopu County

图1-1-2　楼兰男子执杖缂毛织物残
片 Wool Fragments of Loulan Men with
Sticks

　　从材料上看，这些缂毛织物的材料使用羊毛线，由于羊毛纤维短，易于松散和纠缠，因此经线的捻度必须大，同时，上机张力大才能使得经线绷紧；纬线采用弱捻，使纤维蓬松，保持较大程度的弯曲，因此织物呈现纬面凸纹。

In terms of the materials, these wool fabrics were made of wool yarns. Being short and fluffy, wool fibers were easy to loosen and entangle. With low density, the fiber would maintain a larger twisting degree. In the process of weaving, the tension on the machine would make the warp threads tight, while the weft adopted weak twist to make the fibers fluffy, keeping a greater degree of bending. Woven densely, the fabrics presented convex lines on the weft surface.

从结构上看，这些织物既有平纹，也有斜纹；既有纵织，也有横织。背面的色泽、图案与正面相同，但各种色彩的纬线相互穿梭，显得比较零乱。

In terms of the structure, these fabrics had both plain weave and twill weave, vertical weaving and horizontal weaving. The color and the pattern on the back were the same as those on the front, but the wefts of various colors shuttled through each other, which looked rather messy.

从装饰花纹上看，缂毛织物体现当代的审美特征。花纹主要有几何图案、动物图案和植物图案，如回纹、菱纹、锯齿纹、水波纹、变体羊角纹、勾连纹、动物和植物纹、人首马身纹等。

When it came to the decorative patterns, woolen fabrics reflected the modern and contemporary aesthetic characteristics. The decorative patterns included geometric patterns, animal and plant patterns, Such as reciprocal patterns, diamond patterns, zigzag patterns, water ripple patterns, variant goat–horn patterns, hook patterns, animal and plant patterns, and human–head horse–body patterns and so on.

从使用功能上看，我国出土的缂毛以实用品居多，如外衣边饰，裤子边饰，裙带、裙子的腰饰等幅宽较窄的产品，也有缂毛毯子等日用品。

From the functional point of view, most of the woolen fabrics unearthed in China were practical products, such as coat trim, trousers trim, waistband, skirt waist trim, and the width was relatively narrow. And at that time, the woolen fabrics were also used to make daily necessities such as blankets to keep people warm.

二、缂丝的出现与传播/Emergence and Spread of Kossu Fabrics

从出土和传世遗物看，缂丝出现于唐。缂丝出现后慢慢地向着两条路线传播：一条是从西北向西夏、辽、金以及元等北方少数民族传播，主要以实用品为主，后来出现了宗教用装饰画，为"北方传播路线"。另一条是向中原、江南地区传播，受中原和江南地区高度发展的文化影响，从实用品逐渐向装裱用、观赏用缂丝发展，为"南方传播路线"。

缂丝的出现与传播

Judging from the unearthed and handed–down relics, the kossu fabric appeared in the Tang Dynasty. After the emergence of the kossu fabric, it gradually spread to other areas in two routes: one was the "Northern Route", spreading from northwest to northern ethnic minorities such as Western Xia, Liao, Jin and Yuan. On this route, practical products were the majority, and later religious decorative paintings appeared. The other route was the "Southern Route", spreading to the Central Plains and Jiangnan region. Influenced by the highly developed culture in the Central Plains and Jiangnan region, the products of kossu fabrics gradually developed from practical ones to ones for mounting and viewing.

缂丝在唐代出现的条件如下。

Conditions for the emergence of kossu fabrics in the Tang Dynasty are as follows.

（1）汉以来从西方输入的缂毛技术和丝织业的发展是缂丝产生的直接条件。缂毛技术在唐代的西北地区已经成熟并广泛流行。

Since the Han Dynasty, the technology of wool weaving imported from the West and the development of silk weaving industry were the direct conditions for the emergence of kossu skill. In the Tang Dynasty, the technique of making woolen fabrics had been mature and widely used in the northwest region of China.

在唐代，随着蚕桑技术的普及和应用范围的扩大，全国丝绸产量大幅度提高，花色品种增多。官府织染手工业分工细致，出现了丝绸集中产地，其中四川的益州、江苏的扬州和河北的定州最著名，江南各州的丝绸品种和数量也在迅速增长。另外，西北地区丝织技术发展迅速，为缂丝的产生创造了条件。新疆的龟兹（今库车）、疏勒（今喀什）、高昌（今吐鲁番西交河故城）、于田（今和田西南）等地形成丝织中心，生产具有地方特色的锦，如丘慈（龟兹）锦等。高昌地区于5世纪中期建立高昌王国，497年，都城高昌城（今吐鲁番市东哈拉和卓堡西南）随着蚕桑技术在西北地区的普及，人们尝试着以丝线代替毛线织出更加精细华丽的织物。高昌地区还向唐朝宫廷供奉丝。西周治所在今高昌县，这里出土了最早的缂丝带。

In the Tang Dynasty, with the popularity of silk weaving technology and its expansion in scope, the national silk output increased substantially, so did the variety of colors and the types of silk. The government weaving and dyeing handicraft industry had a detailed division of labor, and the intensive silk production areas had emerged, among which Yizhou in Sichuan, Yangzhou in Jiangsu and Dingzhou in Hebei were the most famous ones. The variety and the quantity of silk in Jiangnan provinces were also growing rapidly. In addition, the rapid development of silk weaving technology in Northwest China created conditions for the emergence of kossu fabrics. Silk weaving centers had been formed in Xinjiang's Qiuci (now Kuqa), Shule (now Kashgar), Gaochang (now Turpan, the ancient city of Xijiaohe) and Yutian (now southwest of Hotan), producing brocades with local flavors, such as Qiuci brocade and so on. The Gaochang Kingdom was established in the mid–5th century in the Gaochang area. And in 497, Gaochang City (now East Halla and southwest of Zhuobao in Turpan) was designated as the capital city. With the popularity of silk weaving technology in northwest China, people tried to produce more delicate and gorgeous fabrics with silk instead of wool. Gaochang area also enshrined silk to the court of the Tang dynasty at that time. The seat of government of the Western Zhou Dynasty was located in today's Gaochang County, where the earliest silk ribbons were unearthed.

（2）唐朝政治稳定、经济繁荣，文化艺术领域也取得了长足的发展，外来文化的输入为缂丝等新工艺的出现创造了条件。

The Tang Dynasty was politically stable, economically prosperous and culturally well-developed. The introduction of foreign culture created favorable conditions for the emergence of new crafts such as the kossu technique.

（3）唐朝对外交流空前活跃，与粟特、萨珊波斯、拜占庭等西方国家和东方的新罗、日本建立了广泛的外交关系。各国使者、留学生、商人以及学者纷纷涌入唐朝，把本国文化带到这里。各国的特种手工艺源源不断地传入，经济实力增强、文化的繁荣和西方文化的传入，使统治阶层的物质要求空前提高。国内外的奇珍异宝成为宫廷和贵族猎取的主要目标，这种需求刺激了各种高档手工艺品的产生，使唐代手工业在工艺、品种上取得了飞跃性发展，在装饰纹样上体现出东西方文化的共融性。

The Tang Dynasty was unprecedentedly active in foreign exchanges and established extensive diplomatic relations with western countries such as Sogdian, Sassanian Persia and Byzantium, as well as Silla and Japan in the East. Envoys, overseas students, businessmen and scholars from all over the world poured into the Tang Dynasty to bring their own culture here. With the continuous introduction of special handicrafts from various countries, the enhancement of economic strength, the prosperity of culture and the introduction of western culture had made the material demands of the ruling class unprecedentedly higher. Rare and exotic treasures at home and abroad had become the major pursuits of the court and aristocrats. This demand stimulated the emergence of various high-grade handicrafts, which enabled the handicraft industry in the Tang Dynasty to achieve rapid development in crafts and varieties. The decorative patterns also reflected the integration of eastern and western cultures.

（4）唐朝佛教兴盛，政府和民间投入大量财物修建佛寺，经卷、经袱、佛幡等佛教用品使用珍贵的新兴材料，缂丝即是典型。

Buddhism flourished in the Tang Dynasty. The government and the people invested a lot of money to build Buddhist temples. Buddhist articles such as scriptures, scripture packages and Buddhist flags were made of precious new materials, among which kossu fabrics was the typical one.

（5）佛教于汉代通过丝绸之路传入内地，丝绸之路沿线的高昌、回鹘、吐蕃、西夏都盛行佛教，这个地区也是缂丝的发源地。在以佛教为至高无上信仰的地区，缂丝这种新鲜珍奇的材料首先用于佛教用物。大英博物馆藏敦煌发现的缂丝幡首和几件窄幅带、日本正仓院藏缂丝带均是佛教用物。

Buddhism was introduced to China's inland through the Silk Road in the Han Dynasty. Buddhism prevailed in Gaochang, Uighur, Tubo and Kingdom of Xia along the Silk Road, and this area was also the birthplace of kossu fabrics. In areas where Buddhism was the supreme belief, the kossu fabric, a fresh and rare material, was first used for Buddhist articles. The Buddha kossu flag-heads and several narrow belts found in Dunhuang but collected by the British Museum, as well as the kossu ribbons collected by the Temple Todaiji of Japan, were all supposed to be Buddhist objects.

三、唐代缂丝的特点/Characteristics of Kossu Fabrics in the Tang Dynasty

唐代缂丝以实用品为主，后来出现了宗教用装饰画，如装饰腰带、佛幡首。由于织造技术的原因，当时织造5cm以内的窄幅缂带。这里介绍两件典型的唐代缂丝。

In the Tang Dynasty, kossu fabrics were mainly used for practical purposes. Later, religious decorative paintings appeared, such as decorative belts and Buddha's flag-heads. Due to the limitation of weaving technology, only narrow ribbons within 5 cm were woven at that time. Here are two typical pieces of kossu fabrics in the Tang Dynasty.

第一件是1973年在阿斯塔那张雄夫妇合葬墓出土的几何纹缂丝带，如图1-1-3所示。宽1cm，被剪成长为9.5cm的一段用作女舞俑的束腰带。织物密度为经线15根/cm、纬线116根/cm，织造技术明显高于缂毛。这件缂丝带以草绿作地，以大红、橘黄、海蓝、天青、白色、沉香等八彩织成四叶形图案，此件文物目前馆藏在新疆维吾尔自治区博物馆。

The first item is the kossu ribbon with geometric patterns unearthed from the joint tomb of Zhang Xiong couple in Astana cemetery in 1973, as shown in Figure 1-1-3. A section of 1 cm wide and 9.5 cm long was cut to be used as a girdle for female dancing figurines. The density of the fabric was 15 warps/cm and 116 wefts/cm, and the weaving technology was significantly higher than that of woolen fabrics. It was a four-leaf pattern, with grass green as the ground, woven in eight colors such as bright red, orange, sea blue, azure, white and agarwood etc. This cultural relic is currently collected in the Museum of Xinjiang Uygur Autonomous Region.

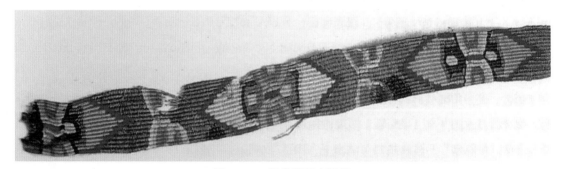

图1-1-3　唐代缂丝束腰带
Kossu Girdle of the Tang Dynasty

第二件是出土于敦煌的白地宝花立鸟缂丝带，如图1-1-4所示，目前收藏在法国吉美博物馆，年代为盛唐—中唐（8世纪）；长56.5cm、宽3.3cm；经线为S捻本色桑蚕丝，密度为18根/cm；纬线为无捻白、绿、黄、橘红、褐、蓝等色的桑蚕丝线，密度约为47根/cm。这件文物图案中心为一宝花中的立鸟纹样，宝花上装饰有四片花瓣及四朵花蕾，均以绿色为主织成，但花瓣中心及花蕾边缘以片金线缂织而成，可以看出片金的金箔已基本脱落，但背衬上有着暗红色的黏合层，局部黏合层中露出背衬底部，应为纸质。团窠采用二二错排的形式，各团窠中的立鸟朝向不同，较为随机。但不同团窠中立鸟及其背景颜色常有变化，其中保存较为鲜艳的是一个红地白鸟，其鸟喙和鸟足为蓝色，翅膀为橙色，整个宝花团窠的大小为经向4.8cm，纬向3cm。从保存的实物尺寸来看，这件缂丝很有可能曾被用作经帙上的装饰带。用缂丝作为装饰带的实例有大英博物馆所藏团窠尖瓣对狮纹锦缘经帙和吉美博物馆所藏团窠尖瓣对狮纹锦缘经帙、联珠对兽纹锦缘经帙。

The second piece is the kossu belt named birds on white treasure flowers unearthed in Dunhuang, as shown in Figure 1-1-4, which is currently collected in the French Musée Guimet, dating from the prime of the Tang Dynasty to the middle Tang Dynasty (8th century). It was 56.5 cm long and 3.3 cm wide. The warp thread was S-twisted natural mulberry silk with a density of 18 threads/cm, while the weft thread was an untwisted white, green, yellow, orange, brown or blue mulberry silk thread with a density of about 47 threads/cm. The central pattern of this cultural relic was a standing bird in a treasure flower. The treasured flower was ornamented with four petals and four buds, all of which were mainly woven in green, but the center of the petals and the edges of the flower buds were woven with pieces of gold threads. It could be seen that the gold foil of the gold piece had almost fallen off, but there was a dark red adhesive layer on the backing, and the backing exposed in the partial adhesive layer should be paper. The cluster adopts the form of two-two staggered arrangement, and the standing birds in each cluster were randomly facing different directions. However, the colors of standing birds and their background often varied in different clusters. Among them, the more vividly preserved was a white bird on red ground, with blue beak and feet and orange wings. The size of the whole treasure flower cluster was 4.8 cm in longitude and 3cm in latitude. Judging from the physical size of the preserved, this kossu belt was likely to be a decorative belt on the sutra. Examples of using kossu fabrics as decorative belts include Cusp-to-lion Cluster Brocade Belt collected by the British Museum, and the Cusp-to-lion Cluster Brocade Belt and bead-to-beast Brocade Belt collected by the Musée Guimet.

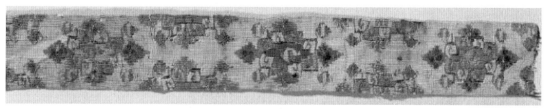

图1-1-4　白地宝花立鸟缂丝带
Kossu Belt with the Design of Birds on White Treasure Flowers

唐代的缂丝具有以下特点：

The kossu fabrics in the Tang Dynasty share some features in the following aspects.

（一）材料/Materials

材料方面，盛唐时期以丝线代替之前的毛线，除了丝线外，还流行片金线缂织，用于勾缂轮廓，强调立体效果。

In terms of the materials, silk thread was used to replace the previous wool in the prime of the Tang Dynasty. Besides silk threads, gold thread weaving was also popularly used to hook the outline and highlight the three-dimensional effect.

片金线出现在唐代，制作流程复杂：首先将金融化，凝为片状，将金片捶打成金叶，使之变硬，经退火处理后，再垂打成金箔，然后在羊皮或纸上刷上鱼胶，粘贴上金箔，并根据不同的要求切成粗细不等的金线。或者在棉纱、蚕丝上涂上胶水，用片金线缠绕搓捻于外围，使之成为圆金线。

The gold thread appeared in the Tang Dynasty, and the production process was complicated: firstly, melted the gold and condensed it into sheets, and then beat the gold sheets into gold leaves to make them harden, and after annealing heat, beat them into gold foils. Next, brushed fish glue on sheepskin or paper and pasted the gold foils on it. And finally, cut them into gold threads with different thicknesses according to different requirements. Or coat glue on a cotton yarn or a silk thread, and then twist a piece of gold thread around it so as to make it into a rounded gold thread.

（二）技法工艺/Techniques

唐代缂丝的主要技法有平缂、掼、构（又称勾边线）、搭梭等。

The main techniques of kossu fabrics in the Tang Dynasty included: flat weaving, Guan weaving, structure weaving (also known as hooking), shuttling weaving and so on.

（1）平缂：用有色的丝线作为纬线与经线交织，按照花纹的色彩要求进行缂织。通常用于作背景底色或小型花纹。

Flat weaving: according to the color requirements of patterns, use colored silk threads as the weft to interweave with the warp threads. This technique is usually employed to create background color or small patterns.

（2）掼：把两种或两种以上相邻的不同颜色根据花纹轮廓走向顺序缂织的方法。

Guan weaving: a method of weaving two or more adjacent but different colors in sequence according to the outline of the pattern.

（3）构：又称勾缂，用与花纹颜色不一致的丝线（通常颜色较深）勾勒出图案的花纹。

Structure weaving, also known as hooking: outline the patterns with silk threads (usually darker in color) that are inconsistent with the color of the pattern.

（4）搭梭：在两种不同颜色的花纹边缘碰到垂直线时，由于两色小梭互不相接，因而留有断痕。搭梭就是为了弥补这种断痕裂缝的补救办法。在断缝每隔一定的距离，将断缝两边的小梭相互来回一搭，绕过对方花纹区内的一根经线，以免竖缝过长，形成破口。

Shuttling weaving: when the edges of two patterns of different colors meet the vertical lines, fractures may occur as two small colored shuttles do not connect with each other. Shuttling weaving is a remedy to make up for this kind of fractures. At intervals of a certain distance between the fractures, the small shuttles on both sides of the broken seams are built back and forth to bypass a warp thread in the pattern area of the other side, so as to avoid the vertical seam being too long to form a breach.

（三）色彩及图案特征/Color and Pattern Characteristics

唐代的缂丝，其色彩主要是平涂的块面，尚未使用退晕和晕染效果，但已经使用金线作底纹。织造纹样多见小的花卉，早期为几何形式排列的图案化的宝相花，盛期出现了写实的折枝花和复杂的联珠纹，唐代墓葬发现花瓣纹缂丝。色彩使用红、黄、蓝、绿等明快的原色系列，柔和的中间色较少使用，因此色彩的跨度大。

In the Tang Dynasty, the colors of kossu fabrics were mainly flat-coated block surface. Though halo removal and halo dyeing had not been used yet, the gold thread had been employed to make shading. Small flowers were often seen in weaving patterns. In the early stage, they were lucky flowers geometrically arranged in the patterns. In the heyday of the Tang Dynasty, realistic folded flowers and complex bead patterns appeared. In the tombs of the Tang Dynasty, kossu fabrics with petals were found. Colors like red, yellow, blue, green and other bright primary colors were more often applied while soft intermediate colors were less used. And that's why the color span was quite large.

◎ 思考题/Questions for Discussion

1. 我国出土的缂毛织物有哪些特点？/What are the characteristics of Kossu wool fabrics unearthed in China?

2. 唐代产生缂丝的条件有哪些？/What are the conditions for the emergence of kossu fabrics in the Tang Dynasty?

3. 唐代缂丝主要技法有哪些？/What are the main techniques of kossu in the Tang Dynasty?

4. 唐代缂丝所用的材料有哪些特点？/What are the characteristics of the materials of kossu fabrics in the Tang Dynasty?

第二节　宋代缂丝/Kossu Fabrics in the Song Dynasty

一、产生的背景/Background

宋代缂丝

北宋时期缂丝逐渐向中原、江南地区延伸，在宋代文化中心——北宋都城汴梁和南宋的临安、苏松地区，缂丝工艺与文化艺术融合，开创了纯艺术性的缂丝，并迎来了缂丝艺术史上的辉煌时期。这就是第一节中提到的"南方传播路线"，是向中原、江南地区传播，受中原和江南地区高度发展的文化影响，从实用品逐渐向装裱用、观赏用缂丝发展，并形成不同的风格。

In the Northern Song Dynasty, kossu fabrics gradually spread to the Central Plains and Jiangnan areas. In the cultural center of the Song Dynasty such as Bianliang, the capital of the Northern Song Dynasty, as well as Lin'an and Susong areas of the Southern Song Dynasty, kossu craft merged with culture and art, creating pure artistic kossu fabrics and having ushered in a glorious period in the history of kossu art. This was the "Southern Route" mentioned in Section 1, by which kossu fabrics spread to the Central Plains and Jiangnan regions. Influenced by the highly developed culture in the Central Plains and Jiangnan regions, kossu fabrics gradually developed from the practical products to the ones for mounting and viewing purposes, and formed different styles.

宋代政权统一，实行"崇文抑武"的政策，经济、文化都得到很好的发展，为缂丝的繁荣创造了适宜的条件。官府设立专门机构管理手工业生产，规模庞大，组织严密。少府监下辖文思院、绫锦院、染院、裁造院等，其中文思院有42个作坊，其中有"缂丝作"27个。另外，内侍省里的后苑造作所有81个作坊，其中也有"缂丝作"。民间手工业异常活跃，家庭作坊从数量和规模上远远超过前代，缂丝的商业化倾向日益明显，在这种环境中缂丝也迎来了辉煌的时代。

In the Song Dynasty, the political power was unified. By implementing the policy of "emphasizing intellectual pursuits and despising martial arts", the economy and the culture were well-developed, which created suitable conditions for the prosperity of kossu fabrics. The government set up a special agency to manage handicraft production, which was large in scale and well-organized. The Directorate for Imperial Manufactories (an agency of the central government supervising a variety of artisan workshops) administered Faculty of Arts and crafts, Silk and Brocade Academy, Dyeing Academy, Tailoring Academy, etc., among which Faculty of Arts and crafts had 42 workshops, including 27 "Kossu Fabric Workshops". In addition, there were totally 81 Houyuan Palace Workshops, a division of Palace Eunuch Service Agency, among which there were also "Kossu Fabric Workshops". As the folk handicraft industry became extraordinarily prosperous and the number and scale of family workshops had far exceeded those of the previous generations, kossu fabrics were becoming more and more commercialized. Under such circumstance, kossu fabrics ushered in the most glorious era.

宋代在文化艺术领域取得了前所未有的辉煌成就，其中绘画艺术首屈一指，对当时缂丝的发展方向产生了重要的影响。宋代绘画题材与唐代有很大的不同，唐代宗教兴盛，宗教画是绘画中的主流，宋代由于经济发展、商业繁荣、科学进步，意识形态领域也发生了很大的变化，宗教观念相对淡薄，因此表现自然景物的山水、花鸟画在北宋逐步取代了宗教、人物画，成为当时人们喜闻乐见的题材，并取得了很大成就。例如，宋徽宗能诗、善书，尤其擅长绘画，人物、花鸟、山水皆为其所长，尤其喜欢花鸟画。宋代崇尚以黄筌为代表的富贵画风，这种端庄富丽、细腻逼真的绘画，非常符合帝王权贵们的欣赏口味和宫廷壁画装饰的需要。随着缂丝、刺绣等技艺的不断提高，人们尝试把名家名作用缂丝、刺绣工艺再现出来，既保存了绘画的传统，又融入了高超的手工技艺。

The Song Dynasty made unprecedented brilliant achievements in the field of culture and art, among which painting art stood out, and exerted a great impact on the development of kossu fabrics at that time. The themes of painting in the Song Dynasty were quite different from those in the Tang Dynasty. Religion flourished in the Tang Dynasty, and religious paintings were the mainstream of paintings. In the Song Dynasty, due to economic development, commercial prosperity and scientific progress, great changes had taken place in the ideological field. People had little religious ideas. Therefore, landscape paintings and flower-and-bird paintings, which showed natural scenery, gradually replaced religion and figure paintings in the Northern Song Dynasty. And these themes became popular subjects at that time, and achieved great success. Forexample, emperor Huizong of the Song Dynasty was good at writing poems and handwriting, especially good at painting. Figures paintings, flower-and-bird paintings, and landscape paintings, all of which were his strengths, especially flower-and-bird paintings. People advocated the rich painting style represented by Huang Quan, and the rich and magnificent, delicate and lifelike paintings would cater for the tastes of emperors and dignitaries and the needs of palace mural decoration. With the constant improvement of the techniques of kossu fabrics and embroidery, people tried to reproduce the famous paintings by using kossu technique and embroidery technique, which not only preserved the tradition of painting, but also incorporated superb craftsmanship.

南宋时期政治、经济和文化中心完全转移到江南，长江三角洲地区汇集了大批文人、画家和工艺美术家。缂丝的生产中心集中在太湖流域，尤其是苏松地区，这里人口密集，商业繁荣，文化底蕴深厚，是缂丝传播和繁荣的肥沃土壤。南宋缂丝在北宋的基础上达到了高峰，缂丝生产极其繁荣。

In the Southern Song Dynasty, the political, economic and cultural center was completely transferred to the south of the Yangtze River, and a large number of literati, painters and crafts artists gathered around the Yangtze River Delta region. The production center of kossu fabrics was concentrated in Taihu Lake Basin, especially in the Susong area. This area was densely populated, prosperous in business and profound in cultural heritage, which provided fertile soil for the spread and prosperity of kossu fabrics. On the basis of the kossu fabrics in Northern Song Dynasty, the kossu fabrics in the Southern Song Dynasty reached their peak. Particularly, the production of kossu fabrics was flourishing at that time.

二、产品特点 /Characteristics of Products

宋代绘画和书法艺术取得卓越的成就，并影响着刺绣、缂丝等工艺品。北宋时以装裱用缂丝为主，装饰花纹表现出图案化、对称化的风格，与当时的装饰艺术风格相符。北宋后期受当时写实绘画艺术的影响，出现了精工摹缂绘画作品的观赏性缂丝，为了真实地再现原画精神，艺人们在缂织技术上不断摸索和创新，使缂织技艺迅速提高。

Painting and calligraphy in the Song Dynasty had made outstanding achievements and exerted a great influence on embroidery, kossu fabrics and other handicrafts. In the Northern Song Dynasty, kossu fabrics were mainly used for mounting.The decorative patterns presented a patterned and symmetrical style, which was consistent with the decorative art style of that time. Influenced by the realistic painting of the late Northern Song Dynasty, ornamental kossu fabrics of fine craftsmanship painting began to appear. In order to truly reproduce the spirit of the original paintings, artists constantly explored and innovated in the kossu weaving technology, which helped to improve the kossu weaving skills rapidly.

南宋时期江东地区成为缂丝生产中心，深厚的文化底蕴、发达的手工业和商业把缂丝工艺推向顶峰。南宋时期以书画为蓝本制作的观赏用缂丝的制作达到巅峰，在数量和技术上远远超过北宋时期。作品更富艺术表现力，能把原作的绘画精神表现得更加准确、完美，达到绘画般的效果。

During the Southern Song Dynasty, the Eastern Yangzte River became the production center of kossu fabrics. The profound cultural heritage, developed handicraft industry and commerce pushed the kossu craftsmanship to the peak. In the Southern Song Dynasty, the production of ornamental kossu fabrics based on calligraphy and painting reached its peak, far exceeding that of the Northern Song Dynasty in quantity and technology. The works were more artistically expressive, and capable to express the original painting spirit more accurately and perfectly to achieve a painting-like effect.

（一）材料 /Materials

"宋缂丝"用丝粗实，捻度强劲，织面挺括而坚实，设色朴质而淳厚，写实与画意并重。在宋代，无论实用性还是观赏性缂丝的经线大多为双股强捻线，这样经线更加粗厚，以致经纬线全部织成后，在经线之间出现纵向的凸起痕迹，俗称"瓦楞地痕"。

"Song Kossu" used thick and solid silk with a strong twist. The weaving surface was neat and solid, and the color was simple and thick, paying equal attention to realism and artistic creation. In the Song Dynasty, the warps of both practical and ornamental kossu fabrics were mostly double-stranded strong-twisting threads, which made the warp threads thicker. So after all the warp and weft threads were woven, longitudinal convex marks could appear between the warps, commonly known as "corrugated ground marks".

（二）技法工艺/Techniques

北宋缂丝，其技法基本上沿袭唐代，为了真实地再现原画精神，艺人们在缂织技术上不断摸索和创新，使缂织技艺迅速提高，花纹比唐代更精细富丽。缂法除唐代的"平缂""掼""构""搭梭"外，又新创了"结"和"参合戗"的方法，作品已具有较高的艺术欣赏性。

In the Northern Song Dynasty, the techniques of kossu fabrics basically followed the Tang Dynasty. In order to truly reproduce the spirit of the original paintings, artists constantly explored and innovated the weaving techniques, which improved the kossu weaving skills rapidly. And the patterns were finer and richer than those in the Tang Dynasty. In addition to the Tang Dynasty's "flat weaving" "Guan weaving" "structure weaving" and "shuttling weaving", the method of "knotting weaving" and "crossing draw weaving" had been newly created. The works had a high artistic appreciation value.

南宋是缂丝工艺的成熟期，其产品逐渐从装饰实用品向艺术欣赏品转移。由于南宋政府提倡绘画，缂丝和刺绣都转向追摹名人名画，以制作纯欣赏性的艺术品为风尚，以缂工极为精细的书画缂丝品为上乘，于是苏松地区出现了朱克柔、沈子蕃、吴煦等缂丝名匠，把以往纺织工艺所不能表现的绘画作品，用手工细致巧妙地缂织出来，使缂丝工艺达到很高的艺术造诣，发展到一个高峰，从这时开始，艺人在作品上创造性地缂织了自己的姓名（印章）。艺人们在不断的实践过程中，不拘泥于固定的技法，根据原作的绘画风格灵活运用手中的梭子，最大限度地发挥了缂丝工艺的特点，摸索新的织造技术和方法，如朱克柔开创的不规则长短戗技法是通过长短不同的各色纬线无规则地参差织入，以色彩的自然变化达到逼真的绘画效果。艺人们为了完成一幅满意的作品，会精益求精地重复制作同一件作品，这种精神把南宋缂丝推向高峰。南宋的技法较之北宋更为成熟、丰富，常见的有：平缂、搭缂、盘梭、长短戗、木梳戗、参合戗、勾缂、子母经等10多种，单幅作品采用的色线也多达十五六种。

The Southern Song Dynasty was the mature period of kossu craftsmanship, and the products gradually shifted from decorative practical products to artistic appreciative ones. As the Southern

Song Dynasty government promoted painting, kossu fabrics and embroidery turned to follow famous paintings of celebrities. The fashion of making purely appreciative works of art, and the fine craftsmanship of painting and calligraphy, made the kossu products superior. Therefore, famous kossu fabrics craftsmen such as Zhu Kerou, Shen Zifan and Wu Xu appeared in the Susong area, who carefully and skillfully wove the paintings that could not be represented by textile craftsmanship in the past, and helped the kossu craftsmanship to a high degree of artistic attainment and developed the craftsmanship to a high level. From then on, artists began to weave their own names (seals) on their works creatively. In the process of constant practice, the craftsmen did not just stick to fixed techniques, but flexibly made use of the shuttles in their hands according to the original painting styles, in an effort to give a full play to kossu skill and explore new weaving techniques and methods. For example, the irregular "long–and–short draw weaving" technique initiated by Zhu Kerou was to weave irregularly and staggered weft threads of various colors and different lengths, to achieve a realistic painting effect through the natural changes of colors. In order to complete a satisfactory work, artists would strive for perfection and repeat the same work. This pushed the Southern Song kossu to the peak. Compared with the kossu skills in the Northern Song Dynasty, the skills of the Southern Song Dynasty were more mature and diversified. There were more than 10 kinds of common techniques, such as flat weaving, tapered weaving, winding shuttling weaving, long–and–short draw weaving, wooden–comb draw weaving, crossing draw weaving, hooking weaving, and Zi Mu warps weaving, etc. What's more, there could be as many as fifteen or sixteen kinds of color lines used in a single work.

（三）纹样/Patterns

北宋时期书法和绘画艺术空前繁荣，并兴起一股收藏热，宫廷收集六朝和唐代的书画作品，并用高级丝织品进行装裱。装饰纹样有花卉、鸟兽、山水楼阁等图案，形式上或对称或交叉排列，还保留着唐代的风格。当时以花鸟为题材的装裱用缂丝相当流行。

In the Northern Song Dynasty, calligraphy and painting flourished unprecedentedly, and there was a craze for collection. The palace collected paintings and calligraphy works of the Six Dynasties and the Tang Dynasty and mounted them with high–quality silk fabrics. Decorative patterns included flowers, birds and animals, landscapes and pavilions, etc., which were arranged symmetrically or crosswise in form and still retained the style of the Tang Dynasty. At that time, it was quite popular to use kossu fabrics with flowers and birds as the theme for mounting.

南宋缂丝除了北宋时期小幅册页外，还出现了大尺寸的立轴。南宋缂丝题材以山水景物、花草鱼虫为多，由宫廷画匠和绘画高手提供画稿，艺人摹稿缂丝追求神似，艺术上达到极高的成就。南宋时期装裱用缂丝的制作更加活跃，装裱用缂丝的纹样构图比较简单，宫廷选用龙凤、楼阁等代表身份和尊严的题材，民间多选用牡丹、寿桃、瑞兽、瑞鸟等吉祥题材。受唐代影响，形式上以对称、交错等固定样式比较多。南宋缂丝除了纯艺术性缂丝产品外，艺人们选择动物、植物中蕴涵的吉祥含义，或神话故事等题材制作祝寿、祝福内容的馈赠佳品，这种题材的作品当时很受欢迎，制作数量逐渐增多。另外，这个时期还有少量佛教、道教题材的作品，这些作品大多数以人物为主题，人物的五官、肌肤、衣褶是最难表现的部分，艺人们用劈丝细线、合花线等材料，体现线条和层次的变化，使人物栩栩如生，惟妙惟肖。

In the Southern Song Dynasty, the major products made of kossu fabrics were flower-and-bird works, as well as the works with other themes such as landscapes and characters. In addition to the small albums from the Northern Song Dynasty, large-sized vertical-hanging paintings appeared. In the Southern Song Dynasty, landscapes, flowers, fish and insects were the main themes of kossu fabric products. Court painters and master painters provided rough sketches first. And then artists copied the manuscripts on the kossu in pursuit of spiritual likeness, and achieved extremely high artistic achievements. The production of kossu fabric for mounting in the Southern Song Dynasty, was more active. The composition of kossu patterns for mounting was relatively simple. The palace preferred to use dragons, phoenixes, pavilions, etc. to represent identity and dignity, while the folks favored to use lucky symbols such as peony, longevity peach, auspicious beast and auspicious bird. In terms of form, there were more fixed styles, such as symmetry and staggering, influenced by the Tang Dynasty. Apart from artistic kossu fabrics in the Southern Song Dynasty, artists chose animals and plants with auspicious connotations, or fairy tales and other themes to make gifts for birthday celebrations and blessings. Works on these subjects were very popular at that time, and the number of works gradually increased. In addition, there were a small number of works on Buddhism and Taoism during this period. Most of these works took characters as the theme. The facial features, skin and clothes folds of characters were the most difficult parts to depict by using kossu skills. Artists used materials such as split silk threads, composite threads and etc., paying attention to changes in lines and layers, so as to make the characters lifelike and vivid.

三、名家名作/Masterpieces

从宋代开始，缂丝有相当一部分由装饰实用品转为独立的艺术欣赏品，其作品精致、细腻、丰富，以梭代织，其书画作品更是如此，出现了很多著名的缂丝艺人和作品。

宋代缂丝代表作

Since the Song Dynasty, a considerable part of kossu products had shifted from practical decorative products to independent artistic appreciation products. Its works were exquisite, delicate and rich, were they were woven by shuttles. So did painting and calligraphy works. Many famous kossu artists and works had emerged since then.

（一）缂丝《紫天鹿》与缂丝《紫鸾鹊谱》/Kossu *God Deer with Purple Background* and Kossu *Phoenixes and Magpie with Purple Background*

缂丝《紫天鹿》是北宋缂丝中非常有名的作品，原为书画包首，馆藏于故宫博物院。作品长45.7cm，宽27.3 cm。该作品以紫色为地，在遍地密花中，间饰天鹿、月兔、异兽、翔鸾纹。其缂织技术并不复杂，主要为平缂和环缂，但在表现手法上却有独到之处。对花叶层次的处理不单借助于渲晕及色彩的配置，而且采用不同的纬线密度来表现，即心部纬线细密，向外逐渐粗疏。彩纬有紫、墨绿、棕黄、橘黄、米黄、淡茶、湖蓝、缥、米色等。织物采用遍地密花的布局以及花枝纹样的造型，用色极富时代特征（图1-2-1）。

Kossu *God Deer with Purple Background* a very famous work in the Northern Song Dynasty, was originally the first painting and calligraphy package collected in the Palace Museum. This work is 45.7 cm long and 27.3 cm wide. This work took purple as the ground in which the flowers were densely scattered everywhere, decorated with sky deer, moon rabbits, alien animals, and flying luan patterns. The kossu weaving technique was not complicated, mainly including flat weaving and ring weaving, but they were unique in technique of expression. The weaving of the mosaic level not only depended on the configuration of shading and color, but also depended on different degree of weft densities. That is to say, the weft threads in the center were fine and delicate and gradually coarse outward. The colored wefts were purple, dark green, brown-yellow, orange-yellow, beige, light tea, lake blue, pale green, beige and so on. The densely-flowered layout of the fabric, the shape and color of the flower branch pattern were all very characteristic of that time（Figure 1-2-1）.

辽宁博物馆馆藏的缂丝《紫鸾鹊谱》与缂丝《紫天鹿》在图案风格上也十分相似。北宋时期书画艺术空前繁荣，带动了装裱用缂丝的发展，其中紫鸾鹊题材非常盛行。现存大多为其残片，此幅完整地保存了两组图案。该作品原件长131.6cm，宽55.6cm，每组由五横排花鸟组成，连续不断的枝叶把牡丹、莲花等花卉连在一起，文鸾、仙鹤、锦鸡、孔雀等禽鸟在花丛中展翅飞翔。唐代风行的团窠对称图案到北宋时期变成横向对称或连续排列的形式，仍保留着明显的装饰意味，与南宋时期自然写实的风格形成对比。此幅以紫色为地，使用蓝、绿、白、黄等深浅不同的色彩表现花鸟的线条层次。用搭梭技法避免裂缝的出现。1920年春天朱启钤游北京琉璃厂海王村，于博韫斋购得此幅，并得知此物出于潍县陈氏，原题为《鸾封奕叶图》(图1-2-2)。

The Kossu *Phoenixes and Magpie with Purple Background* and the Kossu *God Deer with Purple Background* collected in Liaoning Museum were also very similar in the style of patterns. The unprecedented prosperity of calligraphy and painting in the Northern Song Dynasty led to the development of kossu fabrics for mounting, among which the theme of purple magpie was very popular. Most of the existing works of this theme were fragments, but there was one completely preserved with two sets of patterns. The original work was 131.6 cm long and 55.6 cm wide. Each set of patterns consisted of five horizontal rows of flowers and birds. Continuous branches and leaves connected flowers such as peony, lotus and other flowers together. Birds such as phoenixes, cranes, golden pheasants and peacock spread their wings and fly among the flowers. The symmetrical pattern of a cluster that were popular in the Tang Dynasty, became horizontally symmetrical or continuously arranged in the Northern Song Dynasty, but still retaining obvious decorative function, in contrast with the natural and realistic style in the Southern Song Dynasty. This piece of work took purple as the ground color, and used different shades of blue, green, white and yellow to reflect the line levels of flowers and birds. The shuttling weaving technique was used to avoid cracks. In the spring of 1920, Zhu Qiqian visited Haiwang Village in Liulichang, Beijing. He bought this painting from Boyunzhai and learned that it was from the Chen family in Weixian County, and entitled *Phoenixes Flying Through Flowers and Leaves* (Figure 1-2-2).

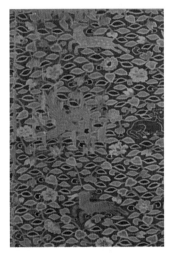

图1-2-1 缂丝《紫天鹿》
Kossu *God Deer with Purple Background*

图1-2-2 缂丝《紫鸾鹊谱》
Kossu *Phoenixes and Magpie with Purple Background*

（二）缂丝《花鸟图轴》与《木槿花图》/Kossu *Flower-and-Bird* and *Hibiscus Flowers*

缂丝《花鸟图轴》馆藏于故宫博物院，长26cm，宽24cm。图轴以宋徽宗赵佶的画稿为粉本进行摹缂。采用平缂、搭缂、盘梭、长短戗、木梳戗、合色线等繁复的技法将花叶的晕色、鸟羽的纹理表现得惟妙惟肖，行梭运丝的细巧使得所缂物象线条柔美，色泽鲜丽，较好地表现了原画细腻柔婉、高雅华贵的艺术风格。图边缂葫芦形朱印"御书"和赵佶的"天下一人"印（图1-2-3）。

Kossu *Flower-and-Bird* collected in the Palace Museum, was 26 cm long and 24 cm wide. This scroll painting was copied from the paintings of Zhao Ji, the Emperor Song Huizong. Complicated techniques such as flat weaving, tapered weaving, winding shuttling weaving, long-

and–short draw weaving, wooden–comb draw weavedraw, and colored–line combining weaving were used to vividly show the halo of flowers and leaves and the texture of bird feathers. The delicate weaving silk made the weaving images soft and bright, and better showed the delicate, soft, elegant and luxurious artistic style of the original painting. The gourd–shaped red seal "Imperial Book" and Zhao Ji's "One Person in the World" were printed on the edge of the picture （Figure 1–2–3）.

辽宁博物馆藏的北宋缂丝《木槿花图》是用缂丝模仿宋徽宗的御制书画，以黄、绿色系为作品主调，缂织折枝木槿花，按照花叶生长的方向退晕缂织。花瓣与绿叶则采用合花线织出色阶的变化，花叶的勾勒线也是断断续续，表现出织者的独特创意。右上方缂织"御书"葫芦印，上有"天下一人"墨押（图1–2–4）。

The Northern Song Dynasty kossu *Hibiscus Flowers* collected in Liaoning Museum, as shown in Figure 1–2–4, imitated Emperor Huizong's imperial paintings and calligraphy by using kossu skills. It took yellow and green as the main colors of the work. The craftsmen weaved hibiscus flowers with folded branches, with the color retreating from the growth direction of flowers and leaves. Petals and green leaves were woven with flower lines to show changes of the color scale, while the outline lines of the flowers and leaves were intermittent, showing the unique creativity of weavers. On the upper right corner, the gourd–shaped seal of "Imperial Book" was woven with the ink "One Person in the World" on it（Figure 1–2–4）.

图 1–2–3　缂丝《花鸟图轴》
Kossu *Flower-and-Bird*

图 1–2–4　缂丝《木槿花图》
Kossu *Hibiscus Flowers*

（三）朱克柔及其作品/Zhu Kerou and Her Works

朱克柔，女，南宋高宗时缂丝名艺人，画家。字朱强、朱刚，云涧（今上海松江）人。以缂丝女红闻名于世，其作品成为当时官僚、文人争相购买的对象。其所作缂丝，人物、花鸟均甚精巧，晕色和谐，清新秀丽。《墨缘汇观·名画》中有明代文人文彦可题字："朱克柔，以女红行世，人物、树石、花鸟，栩栩如生，工品价高，一时流传至今，尤成为罕赠。此尺帧，古澹清雅，有胜国诸名家风韵，洗去脂粉，至其运丝如运笔，是绝技，非今人所得梦见也，宜宝之。"朱克柔熟谙唐以来缂丝技法，并新创了长短戗技法，称"朱缂法"，把缂丝技艺推向了高峰。传世缂丝作品有《莲塘乳鸭图》（上海博物馆藏）、《山茶蛱蝶图》和《牡丹》（辽宁博物馆藏）、《山雀图》《鹡鸰红蓼》《花鸟》《梅花画眉》（台北故宫博物院）等作品，缂工精细，风格高古，形神生动，为南宋缂丝中的代表作。

Zhu Kerou, female, with the courtesy name Zhu Qiang or Zhu Gang, was a famous kossu artist and painter from Yunjian (now Songjiang, Shanghai) in the Gaozong period of the Southern Song Dynasty. She was well-known for her kossu craftsmanship and her works became the objects bureaucrats and literati scrambling to buy at that time. In Kerou's works, figures, flowers and birds were exquisite, and the color was harmonious, fresh and beautiful. In *Mo Yuan Collection—Famous Paintings*, there was the inscription of Wen Yanke of the Ming Dynasty: "Zhu Kerou was famous for her kossu skills. Her works, such as figures, trees, stones, flowers and birds, were delicate and costly, especially as precious gifts, had been passed down to nowadays. Her works, were tranquil and elegant, with the charm winning that of the other famous artists' works. It was the superb skill that by washing away the powder on the draft, and the silk woven in kossu was like a pen in the painting. Such works, only obtained in dreaming, should be cherished and treasured." Zhu Kerou was adept at kossu technique from the Tang Dynasty, and had created a new technique of long-and-short draw weaving method, which, valued as "Zhu's Kossu Skill", pushed the development of kossu technique to a peak. There were 7 masterpieces handed down for generations, including *Suckling Ducks in the Lotus Pond* (collected by the Shanghai Museum), *Camellia Butterflies* and *Peony* (collected by the Liaoning Museum), *Tits*, *Wagtail and Red Polygonum*, *Flowers and Birds* and *Plum Blossom and Thrushes* (collected by Taipei National Palace Museum). They were all representative works of the Southern Song Dynasty with exquisite craftsmanship, elegant style and vivid images.

1. 《莲塘乳鸭图》/Suckling Ducks in the Lotus Pond

如图 1-2-5 所示，此缂丝画幅极大，色彩丰富，丝缕细密适宜，层次分明，是朱克柔缂丝画中的杰作。经近人庞元济收藏，钤"吴兴庞氏珍藏""虚斋秘玩""莱臣审藏真迹"印记。

This kossu painting, large in size, rich in color, delicate in silk patterns, and well-defined in layers, was one of the masterpieces of Zhu Kerou. The seals collected by Pang Yuanji of the contemporary era, were "Wuxing Pang Family Collection" "Rare Playthings of Xuzhai" and "Authentic Works of Lai Chen's Collection"（Figure 1-2-5）.

全幅以彩色丝线缂织而成。图 1-2-5 中双鸭浮游于萍草间，有乳鸭相随，白鹭在侧，翠鸟、红蜻蜓点缀其间。坡岸青石，质感凝重，周围白莲（白藻）、慈菇、荷花（红藻）、萱草等花草环绕，色彩雅丽，线条精谨，生趣盎然。作品中所有花卉虫鸟都极为写实，应以实景写生而成，画风受院体画派影响。图中各种动植物大小体型比例逼真，莲塘的场景时间也能根据乳鸭大小，以及莲塘周围花卉的花期推定为6月底至7月初春末初夏之景。

图1-2-5　缂丝《莲塘乳鸭图》
Kossu *Suckling Ducks in the Lotus Pond*

The whole piece was woven with colored silk threads. In the picture, a pair of ducks were floating among the grass, and there were suckling ducks to follow, egrets on the side, and kingfishers and red dragonflies dotted with them. The bluestone on the slope bank was dignified in texture, surrounded by white lotus (white algae), arrowheads, lotus (red algae), Hemerocallis and other flowers and plants. The elegant color and well-designed lines made the whole picture full of interest. All the flowers, insects and birds in the works were extremely lifelike, and should be sketched from real life. The style of painting was influenced by the school of courtyard painting. The size and build of various animals and plants in the picture was realistic, and according to the size of suckling ducks and the flowering period of flowers around the lotus pond, the scene time of the lotus pond could also be presumed from the end of June to the beginning of July in the late spring and early summer.

图案内容不但丰富且布置合乎庭院自然真实布景：其一，植物都为水生或沼生，喜温暖潮湿环境，而且莲塘和坡地上的植物都属于观赏性的花卉，有荷花（红蕖）、白莲（白蕖）、木芙蓉（芙蓉）、萱草、慈菇、石竹、白百合、芦苇（蒹葭）、玉簪等，应为人工造景；其二，水禽蜓龟也是池塘湖泊常见，有绿头鸭公母一对、乳鸭一对、白鹭一对、燕子一只、翠鸟一只，另有红蜻蜓一只、水黾一对；其三，池塘边站立的以"透、漏、皱、瘦"为美的太湖石，也是就近人工打捞，置于庭院装点。它古朴的青灰色和清奇古怪的形状占据整图的左上角。

Not only was the content of the picture rich and diversified, but also the layout was designed in line with the natural and real scenery of the courtyard. The reasons were as followed: First, the plants were aquatic or marshborn, and they were suitable to live in a warm and humid environment. Moreover, the plants on lotus ponds and on the slopes were ornamental flowers, including lotus (red lotus), white lotus (white vine), hibiscus mutabilis (hibiscus), hemerocallis, arrowheads, carnation, white lily, reed (reeds), hosta, etc. All of these should be artificially landscaped; Second, it was also common to see waterfowls and other animals or insects in ponds and lakes, including a pair of mallard ducks, a pair of suckling ducks, a pair of egrets, a swallow, a kingfisher, a red dragonfly and a water strider; Third, the Taihu Stones standing by the pond were the symbols of beauty, denoting "transparent, leaky, wrinkled and thin" respectively. They were also manually salvaged nearby and placed in the courtyard for decoration. Their quaint bluish–gray color and strange shape occupied the upper left corner of the whole picture.

整幅图中的花卉或并蒂，或结子（也成双数），禽鸟亦俱成对，包括微小的水黾也是一对（靠近上方的翠鸟、蜻蜓及燕子和靠近右侧浮萍位置的白莲虽为独只独朵，推测是处于裁剪位置而被裁掉了），因此根据整幅图选题，可推测所有动植物都应是成双成对，蕴含夫妻和合、多子多福的吉瑞寓意。在《莲塘乳鸭图》中蓝白花线分为多个层次，有浅蓝和白、蓝和白、深蓝和白、浅蓝和深蓝等，区域纬线密度高达120～140根/cm，相当于每十分之一毫米就要织入一根合花纬线（指两根不同颜色的纱合股而成一根线，比如蓝和白两根纱合股作为一根蓝白花线的纬线，放入一个梭子中）。织入合花线的同时，不同合花线之间还要相互戗缂。湖石的浅色部分，比如白色、浅蓝部分缂织的纬线略粗，区域纬线密度稍低，在100～120根/cm，这是由于纬线细密而颜色深会在艺术效果上产生退后的效果，纬线粗实而颜色浅则产生前进凸出的效果，如此一来，玲珑石就显得凹凸有致、玲珑剔透，立体感扑面而来。朱克柔在湖石上缂织题款的用意不仅是因为当时流行奇石题名，如图1-2-6所示，更是因为朱克柔缂的这幅缂丝巨作中，最得她心意的就是这块太湖青石的缂丝技艺，因此她并没有按常规题款于空白处，而是留名于青石之上，或许也有暗含"留名青石（史）"的内心抱负。

In the whole picture, it was common to see that a single stalk usually bore two flowers or had seeds (also in even numbers), and the birds were also in pairs, including a pair of tiny water striders (Though near the top there was only a single kingfisher, a dragonfly and as wallow, as well as one white lotus on the right side of the duckweed, the other of the pair was presumed to be cut off due to the edge cutting position). Therefore, according to the topic of the whole picture, it could be inferred that all animals and plants should be in pairs, symbolizing the harmony of husband and wife, and multiple children and more blessings. In *Suckling Ducks in the Lotus Pond*, the blue–white combined threads were divided into multiple levels, including light blue and white, blue and white, dark blue and white, light blue and dark blue, etc. The section density of weft threads was as high as 120–140 threads/cm, which meant every tenth of a millimeter one combined weft thread must be woven into (combined weft threads: referring to the threads that were formed by combining two yarns of different colors, for example, blue and white yarns were combined as a blue–white weft thread put into a shuttle). When weaving the combined threads, different combined threads with different colors should be woven with each other to grab and adjust the color. The light–colored part of the lake stones, such as the white and light blue parts, had slightly thicker wefts, and the section density of the wefts was slightly lower, ranging from 100–120 threads/cm. This was because fine and dense weft threads were dark to produce the artistic effect to recede, while thick and solid weft threads were light to produce forward and protruding effects. In this way, exquisite stones appeared to be concave and convex, exquisite and translucent, and had a three–dimensional effect. Zhu Kerou's intention of weaving inscriptions on lake stones was not only because inscribing names on strange stones was popular at that time, as shown in Figure 1–2–6, but also because in this masterpiece of kossu fabrics, the kossu skill used in this Taihu blue stone was her favourite. Therefore, she did not inscribe in the blank space as usual, but left her name on the bluestone. Perhaps she also had the ambition of "keeping the name on the bluestone (in the history)".

图 1-2-6　湖石上的缂织题款
Kossu Inscription on the Lake Stone

2.《山茶蛱蝶图》/Camellia and Butterflies

如图1-2-7所示,《山茶蛱蝶图》单幅长25.6cm、宽25.3cm,蓝地织盛开的山茶花,一蝴蝶萦绕其间。左下角缂织"朱克柔印"。此作品以宋代花鸟画为稿本织造,以高超的缂织技巧,纯熟的色彩处理方式完美地表现了原画作的艺术特点。用合花线来表现自然的和色效果。枝干用褐色、米黄色合花线,树叶用绿色、黄绿色合花线来表现。虫蚀部用米黄、褐色线织出自然的色彩过渡,生动传神,这也是此幅画中织造难度较高,较为精彩的部分。花萼处用长短戗织出色彩层次,印章部分用搭梭技法,精湛的技巧使其经线边缘整齐,纬线之间的衔接无缝隙。花瓣的部分用笔补染,从花心处向外渲染出花瓣的色彩和层次,营造出浓淡深浅相宜的效果。对幅有明代文人文彦可题赞。此幅清代初期为书画鉴赏家卞永誉(1645—1712年)收藏。雍正、乾隆间归安岐所有,对幅有"仪周珍藏"印。入清宫后著录于《石渠宝笈》。

Camellia and Butterflies, as shown in Figure 1-2-7, was 25.6 cm long and 25.3 cm wide. Camellia blossoms were woven in blue background, in which a butterfly lingers around. On the lower left corner laid in a kossu-woven Print of Zhu Kerou. This work was woven with the Song Dynasty's flower-and-bird paintings as manuscripts, and perfectly displayed the artistic characteristics of the original paintings with superb kossu weaving techniques and skillful processing of color. The combined color threads were used to display natural color effects. The branches were represented by brown and beige compositing threads, and the leaves were represented by green and yellow-green compositing threads. The section eroded by insects was woven by beige and brown threads to make color transitions natural and vivid. And it was also

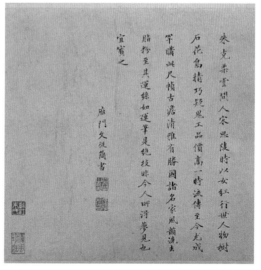

(a)缂丝部分
Part of Kossu

(b)明代文人文彦可题赞
Compliment to the Inscription by Wen Yanke of the Ming Dynasty

图1-2-7 缂丝《山茶蛱蝶图》
Kossu *Camellia and Butterflies*

the more difficult and exciting part of the whole painting. The calyx was woven with long-and short draw weaving to create different color layers. The seal was woven by using shuttling weaving technique. The exquisite technique made the edges of the warp threads neat and the connection between weft threads seamless. The part of the petals was dyed with a pen, and the colors and layers of the petals were rendered outwards from the center of the flower, creating an effect of appropriate shades. Compliment to the inscription was made by Wen Yanke, a Ming Dynasty literary. This painting was collected by Bian Yongyu (1645—1712), a connoisseur of painting and calligraphy in the early Qing Dynasty. During the Yongzheng and Qianlong period, it was owned by Anqi. On the opposite page printed the seal of "Collection of Yizhou". After being collected by the palace of the Qing Dynasty, it was recorded in *Recompilation of Shiqu Treasure Collection*.

（四）沈子蕃及其作品/Shen Zifan and His Works

沈子蕃，南宋缂丝名匠，以摹缂名人书画著称于世，其作品丝纹匀细，工丽典雅，配色古朴，神韵生动。现存作品有5件，即《梅鹊图》《青碧山水图》《秋山诗意图》《山水轴》《桃花双鸟立轴》，无不精妙绝伦。

Shen Zifan, a famous kossu craftsman in the Southern Song Dynasty, was well-known for copying paintings and calligraphy of celebrities. His works were evenly fine-grained, beautiful and elegant in craftsmanship, simple in color matching and vivid in charm. There are five existing works, such as *Plum Blossoms and Magpies in Winter*, *Landscape*, *Mountains in Autumn*, *Scroll of Landscape*, and *vertical scroll of Peach Blossoms and Double Birds*, and all of them are extraordinarily exquisite.

沈子蕃的《梅鹊图》，长104cm，宽36cm。此图轴丝质，依画稿缂织。以十五六种色丝装的小梭代笔，巧妙搭配，画面色泽和谐。以"通经断纬"的手法缂织，并娴熟地运用了多种缂丝技法。所用丝线的经密度为20根/cm，纬密度为44～46根/cm，精工细密。很好地体现了原画稿疏朗古朴的意趣，画面生动，清丽典雅，是沈子蕃为数不多的存世作品之一，也是南宋时期缂丝工艺杰出的代表作（图1-2-8）。

Shen Zifan's *Plum Blossoms and Magpies in Winter* was 104 cm long and 36 cm wide. This scroll was made of silk and woven by kossu techniques according to the original draft. Small silk shuttles with about fifteen colors were skillfully employed to match the picture with harmonious colors. It was mainly woven with the kossu weaving technique of "passing warp and breaking weft", in addition to a variety of other kossu techniques. The warp density was 20 threads/cm, and the weft density was 44-46 threads/cm. It was fine and exquisite in craftsmanship. It well reflected the simplicity and charm of the original painting, vivid, beautiful and elegant. It was one of the few surviving works of Shen Zifan, and an outstanding masterpiece of kossu craftsmanship in the Southern Song Dynasty（Figure 1–2–8）.

沈子蕃的《青碧山水图》长 88.5cm，宽37cm。此图轴采用了织绣中的多种技法，以"参合戗""长短戗"缂成山纹，以"构缂"法织出轮廓勾边线，以"平缂"用于所有的平涂色块，以"子母经"用于缂织文字和图章。此外，在山、云、水等处局部还以淡彩渲染，使景物阴阳远近，层次分明。此件图轴缂丝运梭如运笔，不失分毫，线条勾勒有力，设色明丽天成，它再现了江南大自然空灵开旷的情趣，具有笔墨山水画所不具备的工艺质感之美（图1–2–9）。

Shen Zifan's *Landscape* was 88.5 cm Long and 37 cm wide. This work adopted a variety of techniques in weaving and embroidery. Methods like "crossing draw weaving" and "long–and–short draw weaving" were used to weave mountain patterns; the structure weaving technique was used to outline the picture; the flat weaving was used for all flat color blocks; and Zi Mu warps weaving technique was adopted to weave the words and seals. In addition, mountains, clouds, water and other parts were also rendered in light colors, to make the scenery yin and yang, far and

图1–2–8 缂丝《梅鹊图》
Kossu *Plum Blossoms and Magpies in Winter*

图1–2–9 缂丝《青碧山水图》
Kossu *Landscape*

near, with distinct layers. In this piece of work, kossu shuttles were skillfully operated like a pen, without losing the slightest details. The lines were outlined powerfully, and the colors were bright and beautiful. It reproduced the emptiness and openness of the nature in the area of south of the Yangtze River, and presented the beauty of craftsmanship that could not be possessed by landscape paintings with brush and ink（Figure 1-2-9）.

（五）吴煦及其作品/Wu Xu and His Works

吴煦，字子润，南宋缂丝名艺人，延陵（今江苏常州）人。主要摹缂名人书画，工细巧妙，达到很高的造诣。

Wu Xu, with the courtesy name Zirun, was a famous artist of kossu in the Southern Song Dynasty. He was from Yanling (now Changzhou city, Jiangsu Province). He was good at copying famous paintings and calligraphy of celebrities. With ingenious workmanship, his works achieved high attainments.

《蟠桃花卉图》长71.6cm、宽37.4cm，丰硕的桃实、沧桑的灵芝和奇异的玲珑石构成一幅祝寿画面，山坡间小溪潺潺流过。此幅使用多种技法精工织造：花叶和树干双线勾勒突出轮廓，玲珑石用燕尾戗表现层次，并使用合花线增强写实意味。灵芝环缂出沧桑年轮，立体效果极佳。灵芝、小草、桃子局部稍加补笔。此幅线条柔美，色彩优雅，堪称南宋时期祝寿题材中的代表作（图1-2-10）。

Flat Peaches and Flowers was 71.6 cm long and 37.4 cm wide. Rich peach fruits, ancient ganoderma lucidum and fantastic exquisite stones formed a picture of birthday celebration, with a small stream gurgling through the hillsides. This work was carefully woven with a variety of techniques: flowers, leaves and tree trunks were double-threaded to outline and highlight the shape, exquisite stones were depicted by the dovetails to show the different levels and layers, and the combined threads were used to enhance the authenticity. Ring weaving of the Ganoderma lucidum revealed vicissitudes of life, and the three-dimensional effect was excellent. Ganoderma lucidum, grass, and peaches were partially supplemented. With soft lines and elegant colors, this piece of work could be called a masterpiece of birthday celebrations in the Southern Song Dynasty (Figure 1-2-10).

《蟠桃献寿图》以本色丝为地，用多彩色丝缂织一位仙翁持桃献寿的图案。全幅钤"乾隆御览之宝""乾隆鉴赏""三希堂精鉴玺""宜子孙""嘉庆御览之宝""嘉庆鉴赏""石渠宝笈"和"宝笈三编"，共八玺。画幅上天空中祥云缭绕，红日高照，一只仙鹤凌空飞翔。下方的蟠桃树枝干遒劲，果实累累，树下一位容光焕发、神采奕奕的仙翁手捧刚摘获的一枚硕大蟠桃，欣然回首，健步而行。地面上、寿石旁，灵芝、水仙和翠竹生机益然。在古代，鹤被视为"鹤寿千岁，以极其游"的瑞禽，蟠桃则传为西王母的瑶池中所种植，系三千年一开花、三千年一生实，食一枚可增寿六百年的仙物。二者结合，再配以仙翁寿星合为图案，寓意吉祥长寿，常作寿辰庆贺之用。宋元以来，蟠桃献寿一直是人们非常喜爱的祝寿图案和题材。此幅《蟠桃献寿图》熟练运用平缂、勾缂、掼、结、长短戗和包心戗等多种技法，特别是合色线的使用，更增添了作品的表现力。采用藏青、浅蓝、粉红、月白、淡黄、瓦灰、驼色和墨绿等色丝缂织而成。更为难得的是，这件缂丝作品经《石渠宝笈》著录。《石渠宝笈》共著录了三幅《蟠桃献寿图》，此件应为《石渠宝笈三编·乾清言》所录。据目前公布的资料看，目前国内外各大博物馆尚无一家收藏有此三件缂丝中的任何一件，足见其珍罕程度（图1-2-11）。

Flat Peaches for the Birthday Celebration took natural silk as the ground and used multi-color silk to weave a pattern of an old fairy man holding peaches to offer longevity. There were totally eight seals in this work, including "*Treasure of Qianlong Imperial View*" "*Appreciation of Qianlong*" "*Fine Seal of Sanxi Hall*" "*Appropriate for Descendants*" "*Treasure of Jiaqing Imperial View*" "*Appreciation of Jiaqing*" "*Shiqu Treasure Collection*" and "*Third Edition of the Treasure Collection*". In this work, there were auspicious clouds in the sky, the red sun was shining brightly, and a crane was flying in the air. Below the flat peach branches were dry and full of fruits, and a radiant and vigorous fairy under the tree was holding a huge flat peach that had just been picked, looking back happily and walking with great strides. On the ground, beside the longevity stone, Ganoderma lucidum, narcissus and green bamboo were full of vitality. In ancient times, cranes were regarded as auspicious birds with "a thousand years of life", while flat peaches, which blossomed once in 3,000 years and produced fruits once in 3,000 years, were thought to be planted in the Yaochi of the Queen Mother of the West, and to be fairy fruit to prolong people's life for 600 years by eating one. The combination of the two, together with the design of old fairy man, symbolized auspiciousness and longevity, and it was often used for birthday celebrations. Since the Song and Yuan Dynasties, celebrating birthday with flat peaches had always been a popular birthday pattern and theme for people. This work skillfully adopted a variety of techniques, such as flat weaving, hook weaving, Guan knotting weaving, long-and-short draw weaving, and heart-wrapped draw weaving, especially the use of mixed color threads, which added to the expressiveness of the work. It was woven with silks in dark blue, light blue, pink, moon white, light yellow, tile gray, camel and dark green. What's even rarer was that this kossu work had been recorded in *Shiqu Treasure Collection*. *Shiqu Treasure Collection* has contains three pieces of *Flat Peaches for the Birthday Celebration*, and this piece should be recorded in *Third Edition of Shiqu Treasure Collection·Qian Qing Yan*. According to the currently published information, none of the major museums at home and abroad had any of these three kossu fabrics, which showed how rare they are (Figure 1-2-11).

图 1-2-10　缂丝《蟠桃花卉图》
Kossu *Flat Peaches and Flowers*

图 1-2-11　缂丝《蟠桃献寿图》
Kossu *Flat Peaches for the
Birthday Celebration*

◎ 思考题/Questions for Discussion

1. 北宋和南宋缂丝的花纹有哪些异同点？/What are the similarities and differences between the patterns of kossu fabrics in the Northern Song Dynasty and the Southern Song Dynasty?

2. 宋代具有代表性的缂丝名匠是谁？主要代表作有哪些？/Who are the representative kossu craftsmen in the Song Dynasty? What are the main representative works?

第三节	辽代及西夏缂丝/Kossu Fabrics in the Liao Dynasty and Kingdom of Xia

辽代及西夏缂丝

一、产生的背景/Background

从新疆地区产生和发展起来的缂丝是从西北向西夏、辽、金以及元等北方少数民族地区传播，主要以实用品为主，后来出现了宗教用装饰画，即"北方传播路线"。

Kossu fabrics produced and developed from Xinjiang had spread from northwest areas to northern ethnic minority areas such as the Kingdom of Xia, Liao, Jin and Yuan. And this spreading route was called the "Northern Route". Most of the kossu products at that time were made for practical use. Later, religious decorative paintings appeared.

北方少数民族地区与中原和江南相比，锦绣、绫、罗等高级丝织物的生产能力相对弱。缂丝完全靠手工操作，织机简单，织造方式原始，然而幅宽、花纹可以自由设计，灵活性强，所以比较适合北方少数民族地区生产。在"北方传播路线"上，西夏、辽的缂丝技术达到相当高的水平。

Compared with the Central Plains and Jiangnan, the production capacity of high-grade silk fabrics such as brocades, silk and satin in northern minority areas was relatively weak. Kossu fabrics were completely made by hand. The loom used in making kossu fabrics was simple in structure and the weaving method was primitive. However, the width and pattern of kossu fabrics could be freely designed, so they were more suitable for production in northern minority areas. Along the "Northern Route", the techniques of making kossu fabrics in Kingdom of Xia and Liao had reached a fairly high level.

西夏位于今甘肃、宁夏、陕西北部和内蒙古西部，民族除了羌、藏、汉族外，回鹘人也有相当的数量。西夏与回鹘毗邻，回鹘人发达的纺织技术直接传播到西夏。现存多件西夏缂丝唐卡，如甘肃省黑城遗址出土的缂丝绿度母、美国大都会博物馆藏缂丝唐卡以及西藏拉萨藏品。西夏在绘画、雕刻、石窟寺艺术方面都取得了非凡的成就，被誉为"西夏画派"艺术。西夏盛行佛教，佛教绘画非常发达，有石窟寺院壁画、绢画以及木版画。在敦煌莫高窟、安西榆林窟、安西东千佛洞、旱峡石窟、天梯山石窟等地都有丰富的西夏壁画遗存。这些壁画远承唐、五代，直接受宋、回鹘、西藏、辽的影响，形成自己的特色。高度发达的佛教美术以及良好的艺术氛围对缂丝工艺产生了很大的影响，艺人们具有一定的绘画基础，以优秀的佛画为蓝本，织造具有绘画效果的缂丝作品。

Kingdom of Xia was located in today's Gansu, Ningxia, northern Shaanxi and western Inner Mongolia. In addition to the Qiang, Tibetan, and Han ethnic groups, there were also a considerable number of Uighurs in Kingdom of Xia. Kingdom of Xia was adjacent to Uighurs, and the developed textile technology of Uighurs spread directly to Kingdom of Xia. There were many extant pieces of Kingdom of Xia Kossu Thang-ga, such as Kossu Green Tara unearthed from Heicheng Site in Gansu Province, Kossu Thang-ga collected in the U.S. Metropolitan Museum of Art and Lhasa Collection in Tibet. Kingdom of Xia had made extraordinary achievements in painting, sculpture and cave temple art, and the art style at that time was known as "Kingdom of Xia Painting Style". Buddhism prevailed in Kingdom of Xia, and Buddhist paintings were well developed. Murals in grottoes and monasteries, silk paintings and woodblock prints were quite famous. There were rich relics of Kingdom of Xia mural in Dunhuang Mogao Grottoes, Anxi Yulin Grottoes, Anxi East Thousand-Buddha Caves, Ganxia Grottoes, Tiantishan Grottoes and other places. These murals had been inherited from the Tang and Five Dynasties, and were directly influenced by Song, the murals in Uighurs, Tibet and Liao, forming up their own characteristics. Highly developed Buddhist art and a good artistic atmosphere had a great influence on kossu craftsmanship. Artists with certain foundation in painting, using excellent Buddhist paintings as a blueprint, wove kossu fabrics with a painting effect. The kossu fabrics made by these artists were just like painting themselves.

二、产品特点/Characteristics of Products

从现存实物观察，辽缂丝与西夏缂丝在工艺上有很多相似之处，而与宋代缂丝风格不同。在辽朝，缂丝是皇室和贵族主要使用的纺织材料之一，常用缂丝作为送给宋朝皇室的礼物。

By observing the extant kossu fabrics, the kossu in the Liao Dynasty shared many similarities in technology with the kossu in Kingdom of Xia, but they were different from those in the Song Dynasty. In the Liao Dynasty, kossu fabrics were one of the main textile materials used by royal families and nobles. In Liao, kossu fabrics were also often used as a gift to the royal family of the Song Dynasty.

目前辽墓出土的缂丝数量比较少，主要是1974年辽宁省法库县叶茂台七号墓发现的一批缂丝和缂金，如缂金山龙纹尸衾、缂金佩巾、缂金云水纹软靴、高翅帽等，大多掺入金线缂织。此墓主人为一位契丹贵族的老年妇女，从随葬品和缂金龙纹尸衾看应该与皇室有关。出土的缂丝普遍使用片金线，色彩以黄褐色为主。搭梭、勾缂技术比较熟练，但没有使用色彩退晕法。装饰花纹除了龙，还有花卉、鹿以及人物，以对称形式排列，具有唐代遗风。

At present, the number of kossu fabrics unearthed from Liao tombs is relatively small, among which are a batch of kossu fabrics and kossu gold reels found in Tomb No.7 of Yemaotai, Faku County, Liaoning Province in 1974. For example, the golden kossu dragon-patterned corpse blanket, the golden kossu jeweled scarf, the golden kossu cloud-water pattern soft boots, and high-winged caps, etc. were mainly woven with gold kossu threads. The owner of this tomb was an elderly woman of Khitan nobleman, and from the funerary objects and the golden kossu dragon-patterned corpse blanket, it should be related to the royal family. Gold threads were widely used in unearthed kossu fabrics, and the color was mainly yellowish brown. The techniques of shuttling weaving and hooking weaving were relatively skilled, but the color fading method was not used. In addition to dragons, the decorative patterns also included flowers, deer, and figures, arranged in a symmetrical form, and bore the legacy of the Tang Dynasty.

　　首先，从功能上讲，缂丝都是实用品。主要是袍服、帽子、软靴以及被褥等服装和家庭装饰品。唐代缂丝主要是窄幅装饰带，或做腰带，或用于佛幡的装饰。西夏缂丝也有实用品，但随着佛教的兴盛，逐渐转向缂丝佛像，并取得了突出的成就。宋代，尤其是南宋时期，缂丝主要模仿绘画，以缂丝技法织造出绘画般的纯粹艺术作品，技法逐渐适应绘画的要求，讲究色彩的自然变化和过渡。而辽代缂丝继承唐和西夏的传统，与宋代艺术缂丝的发展道路不同。

　　Firstly, all these kossu fabrics were all functionally practical products, mainly including gowns, hats, soft boots, bedding and other clothing and home decorations. In the Tang Dynasty, kossu fabrics were mainly used as narrow decorative bands, or used as a belt, or used as the decoration of Buddha banners. There were also practical products in Kingdom of Xia kossu fabrics. But with the prosperity of Buddhism, it gradually turned to be used for kossu Buddha statues, which had made outstanding achievements. In the Song Dynasty, especially in the Southern Song Dynasty, kossu fabrics were made mainly by imitating paintings, using the kossu technique to weave pure works of art like paintings. The techniques gradually adapted to the requirements of painting, paying attention to the natural changes and transitions of colors. The kossu fabrics in the Liao Dynasty inherited the tradition of the Tang Dynasty and Kingdom of Xia, but developed differently from that of the Song Dynasty.

　　其次，缂丝在制作工艺上保留着缂毛的简单技法，纬线的走向比较随意，不像宋代缂丝整齐。色彩边缘出现的裂缝用纬线环绕多层，使织物结实耐磨。金线的使用量明显增多，与宋代形成对比，宋缂丝中为了表现绘画效果使用的复杂技法则未见使用。再者，装饰花纹上表现出契丹民族的特色，表现北方自然景色的云纹和山水纹比较多。另外，由于缂丝为皇室和贵族使用的珍奇织物，所以龙凤纹样占很大的比例。

　　Secondly, keeping the kossu wool in the production process was the simple technique of kossu weaving. The weaving direction of the weft was relatively random, unlike the neatly arranged kossu weaving in the Song Dynasty. Cracks at the edge of color were wrapped around by multiple layers of wefts, so as to make the fabric strong and wear-resistant. In contrast with the Song Dynasty, the gold threads were more frequently used, while the complicated techniques used in Song kossu production to reveal painting effects were not used. In addition, the decorative patterns showed the characteristics of Khitan nationality, and cloud patterns and landscape patterns presented the natural scenery of the north. What's more, as the kossu fabrics were rare fabrics used by the royal family and nobles, the dragon and phoenix patterns accounted for a large proportion.

三、名家名作/Masterpieces

　　俄罗斯爱米塔什博物馆藏的《西夏缂丝绿度母》长101cm、宽52.5cm，由甘肃省黑城遗址出土。画面中心为绿度母，上有五坐佛，上下外移部分缠枝莲花之间各有四个跳舞的女子。色彩以蓝、黄两个色系相互对比，如深蓝地上黄色纹样，或黄色地上蓝色纹样，加上勾勒轮廓，使画面主题突出，立体感强。画面中有大小人物，人物的五官、衣褶是织造难度最大的部分，但是都表现得非常形象，缂织技法相当熟练，以勾勒形式强调轮廓，利用搭梭技术防止裂缝的出现（图1-3-1）。

图1-3-1　西夏缂丝《绿度母》
The Kingdom of Xia Kossu
Green Tara

The Kingdom of Xia Kossu *Green Tara* collected by the Hermitage Museum in Russia was 101 cm in length and 52.5 cm in width. It was unearthed from Heicheng Site in Gansu Province. In the center of the picture was a Green Tara, with five sitting Buddhas on it and four dancer girls among the lotus flowers. The colors were contrasted with blue and yellow, such as yellow patterns on the dark blue ground, or blue patterns on the yellow ground. With the outline, the theme of the picture would stand out and present a strong sense of three-dimensionality. There were large and small characters in the picture. The facial features and pleats of the characters were the most difficult parts of weaving, but they were all very vivid. It was quite a skilled knitting technique, with the outline emphasizing profile, and the shuttle weaving technique was used to prevent the occurrences of cracks. As there were close exchanges between the Liao and Kingdom of Xia, the kossu technique was influenced by Kingdom of Xia（Figure 1-3-1）.

◎ 思考题/Questions for Discussion

辽缂丝在功能和制作工艺上有哪些特点？/What are the characteristics of kossu fabrics in the Liao Dynasty in terms of functions and production technology?

第四节　元代缂丝/Kossu Fabrics in the Yuan Dynasty

元代缂丝

一、产生的背景/Background

宋代深厚的文化和艺术底蕴为元、明、清时期文化的发展打下了坚实的基础。元、明、清时期继承了宋代缂丝的传统，并发扬光大，使缂丝艺术持续发展。元世祖忽必烈曾于苏杭设东西织造局，实用性缂丝和观赏用缂丝同步发展，其中实用性缂丝的种类和使用范围有所扩大，如用于服装、靴套、扇子等。缂丝靴套在辽代已有，契丹和蒙古族都是北方游牧民族，缂丝或织锦靴套在当时贵族阶层相当流行，但制作渐趋简单，构图粗犷，色调单纯，体现了统治阶级游牧部的生活习性。元代缂丝还大量用于宗教，史载元世祖忽必烈曾组织缂织了诸天梵像和自己的御容肖像。除了苏松地区外，宫廷和贵族使用的袍服、配饰和日用装饰品缂丝也会在工部、将作院所属织染局等机构生产，各个地方的织染局提举司也织造缂丝。

The profound cultural and artistic heritage of the Song Dynasty laid a solid foundation for the development of Yuan, Ming and Qing cultures. The development of kossu fabrics in the Yuan, Ming and Qing Dynasties inherited the tradition of kossu in the Song Dynasty and carried it forward, so that caused the further development of kossu fabrics. Kublai Khan, the first emperor of the Yuan Dynasty, once set up the East–West Weaving Bureau in Suzhou and Hangzhou, making the practical kossu fabrics and ornamental kossu fabrics develop simultaneously. Among them, the types and applications of practical kossu fabrics had been expanded, such as clothing, boot covers, fans. Kossu boots in the Liao Dynasty had been unearthed. As Khitan and Mongolian were all nomadic nationalities in the north, kossu boots or brocade boots were quite popular among the aristocratic class at that time. But the production of kossu fabrics was quite simple, the composition of pictures was rough, and the color was simple, which reflected the living habits of the ruling class of the nomadic tribes. In the Yuan Dynasty, kossu fabrics were also widely used in religion. According to historical records, Kublai Khan once organized craftsmen to weave the portraits of Buddhist gods and the portrait of himself. Apart from the Susong area, robes, accessories and daily decorations used by the palace and nobles would also be produced in the Ministry of Industry and the Weaving and Dyeing Bureau of the Imperial Manufacturing Institution and other institutions. Kossu fabrics were also supposed to be produced by Supervisorate of the Weaving and Dyeing Bureau in various areas.

二、产品特点 /Characteristics of Products

元代缂丝的发展与唐宋相比，一是本身的图案题材、工艺等有了变化，在元代花卉动物图案受到追捧，特别是牡丹、凤凰一类具有福瑞华贵象征的图案极受欢迎。工艺上阵法细腻，工艺纯熟，具有层次感。蒙古统治者崇尚金色，将金线运用于缂丝，对明清的缂丝发展产生了深远的影响。二是在用途上也灵活多变，缂丝继承宋代成就，实用性缂丝和观赏用缂丝同步发展，其中实用性缂丝的种类和使用范围有所扩大，如服装、靴套、扇子等。

Compared with the kossu fabrics in the Tang and Song Dynasties, the kossu fabrics in the Yuan Dynasty made some changes in the pattern themes and technology. In the Yuan Dynasty, flower and animal patterns were popular, and particularly the patterns used as the symbol of happiness and luxury such as peony and phoenix patterns were immensely favored by people. It was exquisite, skilled and hierarchal in craftsmanship. Mongolian rulers preferred gold and they applied gold threads to the production of kossu fabrics, which had a far-reaching impact on the development of kossu fabrics in the Ming and Qing Dynasties. Secondly, it was also flexible and changeable of kossu fabrics in use. Kossu fabrics in the Yuan Dynasty inherited the achievements of that in the Song Dynasty. The practical kossu fabrics and ornamental kossu fabrics developed simultaneously. Among the kossu fabrics in the Yuan Dynasty, the types and application of the practical ones had been expanded in the aspects such as clothing, boots and fans, etc.

元代缂丝大体分为三类：服用类、宗教类和皇家肖像类。元前中期，蒙古族南下，蒙古族贵族喜好实用性强的服用类缂丝，多用于靴套、腰带、云肩等。因手工业原始，遏制了欣赏性缂丝的发展。元后期，汉文化逐渐影响元代上层阶级的喜好，加之生产水平的提升，出现了欣赏类的作品。元代历史时间短，缂丝数量与宋、明相比少了很多。元代缂丝的生产中心也在苏松地区。

The kossu fabrics in the Yuan Dynasty could be roughly divided into three categories: clothing category, religious category and royal portrait category. In the middle period of the early Yuan Dynasty, Mongolian people went south, and Mongolian aristocrats preferred the clothing kossu fabrics with various practical usages. Kossu fabrics were mostly used for boot covers, belts, cloud shoulders, etc., The primitive handicraft industry at that time restrained the development of appreciative kossu fabrics. In the late Yuan Dynasty, Han culture gradually influenced the preferences of the upper class in the Yuan Dynasty. Together with the improvement of production, works for appreciation began to appear. The Yuan Dynasty had a short history, and the number of kossu fabrics was much less than that of the Song and Ming Dynasties. The production center of kossu fabrics in the Yuan Dynasty was also in the Susong area.

元代缂丝显著的特点是：

（1）元代缂丝已从宋代的纯欣赏品向生活实用品转化。

（2）缂丝大量使用金彩，是缂丝使用金彩的开端。

（3）由于崇尚佛教，以宗教为题材的缂丝品种已较盛行。

The characteristics of kossu fabric in the Yuan Dynasty:

(1) Kossu fabrics in the Yuan Dynasty had been transformed from the purely appreciative products of the Song Dynasty into the practical products for daily life.

(2) The extensive use of gold color on kossu fabrics set the starting point for the use of golden colors in the kossu products.

(3) Due to the popularity of Buddhism, the varieties of kossu products with religious themes

had become more and more popular.

（一）材料/Materials

以宋代雄厚的技术作铺垫，元代的缂织技术也达到了相当高的水准，但与宋代相比有所变化。材料方面，经线开始不加捻或使用弱捻，尤其是在模仿绘画制作的观赏用作品上更加明显，因此织物表面没有瓦楞地痕，织物平整细腻。金线的使用有所增加，用于小面积勾勒轮廓或用于重要部分。元代缂丝极为华丽，风格独特，略显粗犷豪放，缂工依然精细，常用以织造帝后御真画像和佛教绘画。

Taking the strong technology of the Song Dynasty as a foundation, the kossu weaving technology of the Yuan Dynasty had also reached a considerable high level. Compared with the Song Dynasty it had made some changes. In terms of materials, the warp threads were untwisted or weakly twisted, especially in ornamental works. Therefore, the surface of the fabric had no corrugated marks and was smooth and delicate. More gold threads were used to outline the important parts in the small area. It was extremely gorgeous, unique in style, slightly bold and vigorous, but the craftsmanship was still fine. It was often used to weave portraits of the emperor and empress and Buddhist paintings.

（二）技法工艺/Techniques

因统治阶层的喜好，从生产量的角度看，又从欣赏品转向实用品。其特征是用材讲究，追求华丽显示富贵，大量采用真金银丝作纬线，而工艺技法并无发展，艺术效果反而不如宋代。但元代的织金彩，对明清两代的丝织品产生了直接而深远的影响，直到现在，金银彩缂丝品仍很受客商欢迎。

From the perspective of production volume, it had turned from appreciative products to practical products due to the preference of the ruling class. It was characterized by exquisite materials, the pursuit of magnificence to show wealth, and a large number of real gold and silver threads used as wefts, but there was no development in craftsmanship, and the artistic effect was not as good as that in the Song Dynasty. However, the gold–thread weaving of the Yuan Dynasty had a direct and far–reaching impact on the kossu fabrics of the Ming and Qing Dynasties. Until now, the kossu textiles woven with gold and silver threads are very popular.

（三）纹样/Patterns

元代宗教用缂丝数量明显增加，以佛教和道教题材的作品为主。将作院设立"织佛像提举司"，专门织造宫廷佛像，除织锦外还有缂丝。另外，唐卡、经袱用缂丝、织锦等珍贵的材料织成。从道教演化而来的祝寿题材的作品，如以《八仙祝寿》《东方朔偷桃》《赵佶花鸟方轴》等内容织造的缂丝作为雅俗共赏的礼物受到各个阶层人士的青睐。

In the Yuan Dynasty, the production of religious kossu fabrics increased obviously, mainly including Buddhist and Taoist works. The Imperial Manufacturing Institution set up the Supervisor of Buddha Weaving, which specialized in weaving palace Buddha statues, including both kossu weaving and brocade weaving. In addition, Thang-ga (Religious Scroll Painting) and Buddhist sutra wrappers were woven with precious materials such as kossu fabrics and brocade. Birthday congratulation works developed from Taoism, such as *Birthday Congratulation by Eight Immortals*, *Dongfang Shuo Stealing Peach*, *Zhao Ji Flower-Bird Scroll Painting*, were favored by people from all walks of life as gifts of both elegance and vulgarity.

元代风行山水画，山水画不太适合做缂丝的稿本，加上元代开始一改宋代绢地画，而在纸上绘画，纸张比起丝绢来，表现力更强，尤其表现山水画的远近、浓淡以及皴法等笔墨手法更加丰富，要原样织出这种山水画，难度极大。两宋花鸟画追求逼真的效果和瑰丽的色彩，注重线描，这种画风特别适合作缂丝的稿本，因为缂丝是以色彩和线条取胜的。到了元代，在文人画风的影响下，推崇梅、兰、竹、菊等君子题材的写意画，且多水墨，少色彩。没有色彩的写意花卉对缂丝艺术家是没有吸引力的，这是元代观赏性缂丝减少的原因之一。

Landscape paintings were popular in the Yuan Dynasty, but they were not suitable to be manuscripts for kossu products. In addition, people in the Yuan Dynasty began to change the silk painting inherited from the Song Dynasty to paintings on paper that were more expressive than silk paintings. And the brush and ink techniques of the paper paintings were more abundant, especially in the aspects such as distance, lightness, and intensity, etc. And it was extremely difficult to weave such a wonderful landscape painting on silk as it was. Flower-and-bird paintings of the Song Dynasties pursued realistic effects, magnificent colors, and line drawing. This style of painting was particularly suitable for manuscripts of kossu products because kossu textiles won with colors and lines. In the Yuan Dynasty, under the influence of the literati painting style, freehand paintings with gentleman themes such as plum, orchid, bamboo, and chrysanthemum were admired, with more ink and less color. The freehand brushwork of flowers, without any color, was not attractive to kossu artists, which was one of the reasons for the decline of ornamental kossu works in the Yuan Dynasty.

三、名家名作/Masterpieces

（一）宗教类缂丝作品/Religious Kossu Works

元代对宗教采取开放政策，宗教文化走向多元。其中藏族文化的传播对元代产生了极其深远的影响，例如佛教，缂丝也被运用于佛教物品。元代的佛教缂丝作品主要以藏传佛教卷轴画为主，一般尺幅较大，制作精致，在风格上也明显受到汉文化的影响，如现藏于美国纽约大都会博物馆的缂丝《金刚持曼陀罗》，长宽均超过2m，画像精美，细节考究。元代宗教用缂丝数量明显增加，以佛教和道教题材的作品为主（图1-4-1）。

The Yuan Dynasty adopted an open policy to religions, and religious culture became diversified in the Yuan Dynasty. The spread of Tibetan culture, such as Buddhism, had a profound influence on the Yuan Dynasty. And kossu fabric began to be used in the production of Buddhist items. The Buddhist kossu works of the Yuan Dynasty were mainly the Tibetan Buddhist scroll paintings and Thang-ga, which were generally larger in size, exquisite in the craftsmanship and obviously influenced by Han culture in style. For example, the kossu *Vajrayana Mandala,*

collected by the Metropolitan Museum of Art in New York, USA, with a length and width of more than 2m, displayed exquisite portraits and fine details of Buddha images. In the Yuan Dynasty, the production of religious kossu fabrics increased obviously, mainly including Buddhist and Taoist works（Figure 1-4-1）.

图1-4-1　缂丝《金刚持曼陀罗》
Kossu *Vajrayana Mandala*

缂丝《赵佶花鸟方轴》馆藏于故宫博物院，长24.5cm，宽25.3cm。此为宋徽宗赵佶所绘花鸟册页做蓝本的缂丝杰作。画面上，粉红色牡丹或含苞或怒放；立于牡丹枝头的鸟雀俯视着小瓢虫，曲尽其妙；展翅的蝴蝶风姿绰约。整幅画面洋溢着春天的气息。此图缂工精致平细，晕色之处以长短戗、木梳戗、参合戗等技法完成，无着笔。辅以平缂、构缂技法，使画面层次分明，立体感强。用长短戗技法表现的鸟羽、蝶翅等细微之处可谓巧夺天工。摹缂古人书画作品是元代缂丝作品的特点之一。绘画作品以缂丝技法传承和表现，既可传达出中国画的意境和神韵，又与纯粹用笔墨渲染的绘画相异，别有一番情趣。此缂丝图为这类作品的代表作之一（图1-4-2）。

Kossu *Zhao Ji's Flowers and Birds Scroll Painting,* collected by the Palace Museum, was 24.5 cm in length and 25.3 cm in width. This was a masterpiece of kossu based on the album of flowers and birds painted by Zhao Ji, Emperor Huizong of the Song Dynasty. In the picture, pink peonies might be budding or blooming; the bird standing on the peony branch looked down at the little ladybug in a subtly and skillfully way; the butterfly spread its wings gracefully and tactfully. The whole picture was permeated with the breath of spring. This painting was exquisite and thin, with the halo completed by the techniques such as long-and-short draw weaving, wooden-comb draw weaving and crossing draw weaving, instead of using a pen. Supplemented by the techniques of flat weaving and structure weaving, the picture presented a clear hierarchical structure and a strong three-dimensional effect. The subtle details, such as bird feathers and butterfly wings presented by long-short weaving techniques, could be viewed as ingenious and amazing. The works of painting and calligraphy copied by ancient people could be the typical feature of the works of kossu in the Yuan Dynasty. The painting works were inherited and presented by the kossu technique, which not only conveyed the artistic beauty and charm of Chinese painting, but also had a special taste differed from paintings rendered purely with pen and ink. This tapestry picture is one of the masterpieces of this type of work（Figure 1-4-2）.

图 1-4-2 缂丝《赵佶花鸟方轴》
Kossu *Zhao Ji's Flowers and Birds Scroll Painting*

缂丝《八仙庆寿》馆藏于故宫博物院，长 100cm，宽 45cm。此缂丝图以十余种色丝装成的小梭在白色丝地上依画稿缂织出八仙人物及南极仙翁的形象，表现出八仙祝寿的主题。人物面目清晰，神情平和，服饰飘逸。鹤、鹿、流云、山石、修竹等衬景为画面增添了动感和高雅脱俗的仙境氛围。图的创意、构图、缂织技法均具有元代缂丝的特点。其经线直径为 0.15mm，纬线直径为 0.33mm，经向密度为 16~20 根 /cm，纬向密度为 48~50 根 /cm。缂法以长短戗、木梳戗、平缂、搭缂、构缂、参合戗为主，运用了多种技艺。画面线条流畅舒朗，配色典雅，而少量的用金使以冷色调为主的画面平添了几许富丽华美。此缂丝图轴堪称绘画与织造工艺相结合的杰出作品之一（图 1-4-3）。

Kossu *Eight Immortals to Ealebrate Birthday* is collected in the Palace Museum, with a length of 100 cm and a width of 45 cm. The kossu work was made by the small shuttles with more than ten kinds of colored silk threads, with which the image of the Eight Immortals and the Antarctic Fairy were woven on the white silk floor according to the manuscript, showing the theme of birthday celebration by Eight Immortals. The figures had clear faces, peaceful facial expressions, and elegant costumes. Crane, deer, flowing clouds, mountains and rocks, bamboo and other background scenes added a dynamic, elegant and refined fairyland atmosphere to the painting. The creativity, composition and kossu weaving techniques of the work manifested the characteristics of kossu fabric of the Yuan Dynasty. The warp diameter was 0.15 mm and the weft diameter was 0.33 mm; warp density was 16–20 threads per centimeter, and weft density was 48–50 threads per centimeter. The kossu weaving techniques mainly involved were quite diverse, including long–and–short draw weaving, wooden–comb draw weaving, flat weaving, tapered weaving structure weaving and crossing draw weaving. With smooth lines and elegant colors, a small number of gold threads helped to add the cool–toned picture a touch of splendor and magnificence. This artwork is an outstanding combination of painting and weaving craftsmanship（Figure 1–4–3）.

图1-4-3　缂丝《八仙庆寿》
Kossu *Eight Immortals to Celebrate Birthday*

（二）皇家肖像类缂丝作品/Royal Portrait Kossu Works

皇家肖像类缂丝作品特指帝王织御容的缂丝作品，主要用于供奉、祭祀和瞻仰。织御容属于御容的一种，还有一种是绘御容。早在唐宋时期，绘御容就已经出现，但织御容却为元代专有。

Royal portrait kossu works referred to the kossu works of the portraits of the emperor or the empress, mainly used for sacrifice, worship and admiration. Besides weaving of the royal portraits, there was another type of royal portraits—painting of the royal portraits. As early as in the period of the Tang and Song Dynasties, painting of the royal portraits had appeared, but the weaving of the royal portraits was exclusive to the Yuan Dynasty.

◎ 思考题/Questions for Discussion

1．元代缂丝与之前朝代的缂丝相比有哪些显著的特点？/Compared with the kossu fabrics in previous dynasties, what are the characteristics of the kossu fabrics in the Yuan Dynasty?

2．元代缂丝纹样的特点及种类有哪些？/What are the features and types of kossu patterns of the Yuan Dynasty?

第五节　明代缂丝/Kossu Fabrics of the Ming Dynasty

一、产生的背景/Background

明代缂丝艺术取得了一定的成就，使用范围日益扩大，观赏用和实用品的种类大幅增加，织造技术进一步提高。缂丝的生产中心还是在苏州，初期小规模生产，中期开始逐渐兴盛，到万历前后非常繁盛。

明代缂丝

In the Ming Dynasty, the art of kossu had made some achievements. The scope of application was expanding gradually; the varieties of ornamental and practical products were greatly increased; the weaving technology was further improved. The production center of kossu was still in Suzhou: in the early stage, it was a small-scale production; in the middle stage it began to thrive gradually, and it became very prosperous around Wanli years (the reign of Ming Shenzong).

明初朝廷力倡节俭，规定缂丝只能用于敕制和诰命，不能用于各类衣物。明中期以后，社会经济生活已经发展得十分繁荣。明代宣德时期禁令渐弛，缂丝再兴。缂丝品除宫廷御用和官僚达贵享用外，也进入了民间富豪人家。随着商品经济的发展，手工业生产取得了明显的进步，各个部门生产规模不断扩大，产量大幅增加，很多工艺流程和技术得到改良。民营手工业日趋兴旺，中期以后逐渐超过官营手工业，到明代中后期成为手工业生产的主体力量。

In the early Ming Dynasty, the imperial court advocated frugality, stipulating that kossu could only be used for imperial edicts and imperial mandate, not for all kinds of clothes. After the mid-Ming Dynasty, the society and economy developed prosperously. During the Xuande period of the Ming Dynasty, Kossu production revived again as the ban on kossu fabrics was gradually eased. In addition to being enjoyed by the imperial court and bureaucrats, kossu products had also entered the folk rich families. With the development of the commodity economy, handicraft production had made significant progress. The production scale of various departments had been continuously expanded; the output had increased significantly; and many technological processes and technologies had been improved. The private handicraft industry became more and more prosperous. After the middle period, it gradually surpassed the official handicraft industry and became the main force of handicraft production in the middle and late Ming Dynasty.

商品经济的发展冲击了传统的观念。以手工业、商业富裕起来的城市人群成为奢侈品以及文化艺术品的消费者。当时除了收藏古董外，喜好古玩之风流行于江南的富商士绅之间，在全社会形成了一股风气。于是沈周、唐寅、文徵明等人的书画作品成为人们收藏的抢手货。明万历年间，艺术品的生产和需求最旺盛，以绘画为稿本制作的缂丝、刺绣作品也成为流行物。文化的繁荣为缂丝等工艺美术行业注入了新鲜的养料。江南地区自东晋以后文化家族形成，并世代相传，注重文化和教育。继承宋元优秀的绘画传统，苏州形成了著名的"吴门画派"，于嘉靖年间达到鼎盛，为观赏用缂丝的发展提供了良好的素材。总之，苏松地区的经济、文化艺术环境为缂丝的繁荣提供了良好的社会条件。明代家庭手工业是民间手工业生产形式，缂丝就是在家庭手工作坊生产的。从手工艺类型上讲，缂丝属于工艺美术手工业，与丝织手工业不同，大多是分散的个体性质的家庭手工业，处于半农半织的状态，还没有从农业中解脱出来。劳动者运用自己简陋的生产工具从事生产，完全凭借手工操作，生产的技艺在手工业者的眼手之间，大多数产品主要被宫廷、官僚豪绅所占有、消费和享用，所以市场狭小，生产率不高。为了保住市场，艺人们往往严守世代相传的技术秘密。但是在整个社会经济的大趋势下，明代缂丝织造技术比前代有了很大的进步。

The development of commodity economy had exerted a great impact on the traditional notions. Urban populations who had become affluent with handicrafts and commerce had become consumers of luxury goods and cultural works of art. At that time, besides collecting antiques, appreciating antiques was popular among wealthy businessmen and gentry in the south of the Yangtze River, forming a trend in the whole society. As a result, the calligraphy and painting works of Shen Zhou, Tang Yin, Wen Zhengming and others had become sought-after collections among people. During the Wanli period of the Ming Dynasty, the production and demand of artworks were the most vigorous, and made with paintings as manuscripts, kossu and embroidery works became popular. The cultural prosperity injected fresh nourishment into arts and crafts industries such as kossu. Since the Eastern Jin Dynasty, cultural families had been formed in the Jiangnan area and passed down from generation to generation, emphasizing culture and education. In Suzhou, the famous Wu Painting School was formed, inheriting the excellent painting traditions of the Song and Yuan Dynasties. And it reached its peak during the Jiajing period, providing good materials for the development of ornamental kossu fabrics. In short, the economic, cultural and artistic environment of the Susong area provided good social conditions for the prosperity of kossu. In the Ming Dynasty, the family handicraft industry was a form of the folk handicraft production, and kossu fabrics were produced in family-based handicraft workshops. In terms of handicraft types, kossu was a handicraft industry of arts and crafts. Unlike the silk weaving handicraft industry, it was mostly scattered individual household handicrafts, in a state of semi-agricultural and semi-weaving, and had not yet been freed from agriculture. Laborers were engaged in production with their simple and crude tools, solely relied on manual operation. The production skills were in the hands of handicraftsmen. Most of the products were occupied, consumed and enjoyed by the court, bureaucrats and tycoons. Therefore, the kossu market was small and narrow, and the productivity was not high. In order to preserve the market, these craftsmen rigidly kept the technical secrets passed down from generation to generation. However, under the general social and economic trend, kossu weaving technology in the Ming Dynasty had made great progress compared with the previous generation.

二、产品特点/Characteristics of Products

在明代，首先，能够织造大尺寸的缂丝服装，如万历皇帝墓葬出土的衮服，体现了这个时期缂丝技术的较高水平。其次，在材料上金线的使用比较多，帝王服装上使用孔雀羽线，说明缂丝技艺的提高。再次，在观赏性缂丝作品的细微烦琐之处以画代织。明后期开始，手工艺品的商业化倾向日益严重，并波及缂丝行业。经济利益的诱惑，促使应酬之作逐渐增多，可以说这是缂丝艺术衰退的标志。

First of all, in the Ming Dynasty, weaving large-sized kossu costumes were feasible, such as robes worn by the emperor the Wanli Emperor, which reflected the higher level of kossu techniques in that period. Secondly, the gold thread used in materials was more frequent. Peacock feather threads were used in imperial costumes, which was a symbol of the improvement of kossu process. Thirdly, it was to replace weaving with painting in the subtle and cumbersome parts of ornamental kossu works. Since the late Ming Dynasty, the production of handicrafts had become increasingly commercialized, and spread to the kossu industry. The temptation of economic interests promoted a large number of ordinary quality works, which was a sign of the decline of kossu techniques.

（一）材料/Materials

（1）大量缂金。金线有赤圆金及淡圆金（均为捻金线）两种。金线做法，先用纯金锤打成极薄的金叶，再夹在油烟熏炼的"乌金纸"里，继续锤打5~6小时，把金叶打成金箔，交由切箔工人按规定长宽切划整齐，贴于上胶的毛边纸上，称为"擂金"。再将金纸进行磨光磨亮，称为"砑金"。将磨光磨亮后的金纸切成细条金片，把这种金片（即"片金"）直接织于丝织品中，就是"明金法"。如果再将这种金片包捻于丝线上，便是捻金线。用赤金做成的捻金线称"赤圆金"，用黄金制作的捻金线称为"淡圆金"。明代缂丝所用的捻金线，直径细度可达0.2mm，每厘米能缂织捻金线90根以上。这种细密程度也是惊人的。

A number of gold threads. There were two kinds of gold thread: red round gold and light round gold (both of them were twisted gold threads). The gold thread was made by hammering pure gold into extremely thin gold leaves, then sandwiching them in the Black gold paper smoked by oil smoke, and then hammering for 5 or 6 hours until the gold thin sheet became a gold foil, which was handed over to the foil cutter to cut neatly according to the specified length and width, and pasted on the glued rough edge paper, which was called Huo gold. Then the gold paper was polished, which was called Ya gold. The polished gold paper was cut into several thin gold pieces (or Pian gold), which would be woven into kossu fabrics, which was the Ming gold method. The processed gold sheet could be twisted on the thread, which was the twisted gold thread. If the gold pieces were processed and wrapped on the silk thread, it was the twisted gold thread. The twisted gold thread made of red gold was called red round gold, and the twisted gold thread made of gold was called light round gold. The twisted gold thread used in the Ming Dynasty could amazingly reach 0.2 mm in diameter and more than 90 twisted gold threads could be woven by per centimeter of the thread.

（2）采用孔雀羽线缂制花纹。孔雀羽线的做法是从孔雀翎上摘取绒羽，将绒羽一根一根与丝线合并，再以另一根丝将绒羽与丝线缠绕到一起。定陵地下宫殿出土的缂丝龙袍，及故宫博物院藏明代缂丝椅披、桌垫上的龙纹图案，有不少就是采用孔雀羽缂织成花纹的，孔雀羽缂织的花纹金翠夺目，而且很少褪色或变色。

Use peacock feather thread to make patterns of kossu. The practice of peacock feather thread was to take the feather from the peacock feather, and combined feathers with threads accurately, and then wound feathers and silk threads with another thread. Many of the kossu robes worn by the emperor unearthed from Dingling Underground Palace, the kossu chair wrap collected in the Palace Museum and the dragon patterns on table mats were all woven with peacock feathers. The patterns embroidered with peacock feathers were dazzling, rarely fade or change color.

（3）明代缂丝多数采用极细的双股强捻丝线。例如，定陵出土的一件加金、孔雀羽团龙十二章万寿福袍，经线直径0.15mm，经密20根/cm；纬线直径0.2mm，纬密达88～100根/cm。明缂丝，用线细匀，捻度松弛，织面柔软而纤细，设色华丽，装饰与画趣相伴，多用画笔与绣针点缀添色。

Most of the kossu threads in the Ming Dynasty were made of extremely fine dual-strand double-twisting threads. For example, A Twelve Patterns Wanshou Robe with gold and peacock feather unearthed in Dingling had a warp diameter of 0.15 mm and a warp density of 20 pieces per centimeter; the weft thread had a diameter of 0.2 mm and a weft density of 88–100 threads per centimeter. The kossu threads of the Ming Dynasty were refined, flabby in twisting, they touched soft and slender, owning gorgeous colors, accompanied by decorations and paintings. And they embellished the color with brushes and embroidery images.

（二）技法工艺/Techniques

明代前期是缂丝的衰落时期，直到宣德年间才得以复兴，明中晚期的缂丝，其工艺技法承沿南宋，并新创了极富装饰趣味的"凤尾戗"。"凤尾戗"的戗头一粗一细相间排列，粗者短而细者稍长，形似绘画中凤尾的形状，故名为"凤尾戗"。

The early Ming Dynasty was the decline period of kossu, which was not revived until the Xuande period. In the middle and late Ming Dynasty, the kossu was inherited from that of the Southern Song Dynasty. And a new skill called "tail of the phoenix" with great decorative interest. The head of "tail of the phoenix draw weaving" was in a thick and thin arrangement. The thick one was slightly short than the thin one, whose shape was similar to the tail of the phoenix in painting, so it was called phoenix tail draw weaving.

（三）纹样/Patterns

明中晚期缂丝艺人大量采用吴门画派沈周、文徵明、仇英、唐寅等代表人物的画稿为粉本，使缂丝品的身价、品位大大提高，作品具有很高的艺术欣赏价值。明代人物的缂织达到了前所未有的水平，可以摒弃"人物面相以笔绘"，全部用缂丝手法来表现完成。明代观赏性缂丝题材广泛，除了纯艺术的以名人绘画为蓝本的作品外，祝寿、祝夫妻美满、生子以及升官题材的作品逐渐增多。另外，佛教用缂丝唐卡的生产在明代仍然继续，有些僧侣甚至亲自织造。

In the middle and late Ming Dynasty, a large number of kossu artwork artists used the paintings of Shen Zhou, Wen Zhengming, Qiu Ying, Tang Yin and other representatives of Wu Painting School as their manuscripts, which greatly improved the value and taste of kossu products, and these artworks had high artistic appreciation value. The kossu weaving used in characters of the Ming Dynasty reached an unprecedented level, which could use kossu techniques to take the place of drawing faces of the characters. With the addition of color painting, it became more beautiful and elegant. In Ming Dynasty, there were wide-ranging themes of ornamental kossu artworks. Besides pure artworks based on celebrity paintings, there were various works on birthday celebrations wishing couples a happy life, having children and getting a promotion. In addition, the production of Buddhist kossu Thang-ga continued in the Ming Dynasty, and some monks even woven it themselves.

三、名家名作/Masterpieces

（一）十二章福寿如意衮服/Ceremonial Robe of Rulers with Twelve Patterns Good Fortune

在明朝中后期，苏松地区经济持续发展，缂丝生产极其繁荣，万历皇帝陵墓出土缂丝共8种，29件。出土物中以十二章福寿如意衮服最奢华，如图1-5-1所示。衮服的纹样除了十二章、十二团外，龙纹四周装饰八吉祥、祥云等纹饰。另外，遍身还织有279个"卍"字，256个"寿"字，301只蝙蝠和271个如意。在宫黄色地子上用28种色线缂织，大量使用金线、孔雀羽线。出于实用和节俭的目的，明代宫廷多数缂丝服装仅织主要图案，缝缀于服装上。

In the middle and late Ming Dynasty, the economy of Susong areas developed continually, and the production of kossu textiles was prosperous. There were 8 kinds and 29 pieces of kossu textiles unearthed from the mausoleum of Wanli emperor, one of the emperor in the Ming Dynasty. Among the unearthed objects, Ceremonial Robe of Rulers with Twelve Patterns Good Fortune was the most luxurious, as shown in Figure 1-5-1. In addition to twelve chapters and twelve regiments, the dragon pattern was decorated with eight auspicious and auspicious clouds. In addition, there were 279 "卍" characters, 256 "寿（longevity）" characters, 301 bats and 271 Chinese knots. Craftsmen embroidered these images with 28 kinds of colored threads on the yellow ground of the cloth. Gold and peacock feather threads were widely used. Due to practicality and frugality, most kossu costumes in the Ming Dynasty court only was woven with the main images which were fabricated by the kossu process. The rest of the costumes was made by other traditional techniques.

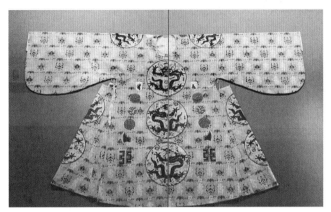

图 1-5-1　缂丝十二章福寿如意衮服
Kossu Ceremonial Robe of Rulers with Twelve Patterns Good Fortune

（二）《瑶池吉庆图》/*Celebration at Jasper Lake*

《瑶池吉庆图》馆藏于故宫博物院，长260cm，宽205cm。此图描绘了神话传说中西王母在西天瑶池庆寿的场景。9位仙女各捧寿礼，凤凰、仙鹤、天鹿、祥云、灵芝、青松、翠柏点缀其间，以强调吉庆祝寿的主题。

Celebration at Jasper Lake is collected in the Palace Museum, with 260 cm in length and 205 cm in width. This picture depicted the scene of the Chinese myth that Queen Mother of the West was celebrating her birthday at Jasper Lake of the Western Heaven. Nine fairies each hold birthday gifts, and phoenix, crane, deer, auspicious clouds,ganoderma, pine and cypress were dotted, which emphasized the theme of auspicious birthday celebration.

此图轴宽两米有余，可见其工程之浩繁，在历代缂丝作品中亦属鲜见。其织缂并未因画幅巨大而粗率逊色，反而越显精妙繁复，行梭运丝巧于变幻，除常见的平缂、搭缂外，细部晕色采用长短戗、木梳戗以及华丽的凤尾戗，仙女鬟角用长短戗技法使发丝自然写实，衣纹用构缂技法突出了飘逸轻柔的感觉，瑶池的水波纹及五彩祥云用掼缂和"结"的技法使之更具立体效果。此图轴画面布局饱满、大气，设色华丽精妙，是庆贺寿辰时的精美装饰品（图1-5-2）。

The width of the handscroll was more than two meters, from which we can see its complex processes. This kossu painting was rare in all dynasties. Although the project was complicated, its weaving process was exquisite. Shuttle–carrying was skillful in changing. In addition to the common flat weaving and structure weaving, long–and–short draw weaving, wooden–comb draw weaving and phoenix tail draw weaving were applied in the bleeding of the details.

图 1-5-2 缂丝《瑶池吉庆图》
Kossu *Celebration at Jasper Lake*

Fairy sideburns were embroidered by long-and-short draw weaving, so their hair was natural and realistic. Structure weaving was used in clothing patterns to highlight the elegant and gentle feeling. Guan weaving and knotting techniques were adopted in the water ripples at Jasper Lake and in the colorful auspicious clouds, with a three-dimensional. The handscroll was on the plump layout. Its color was gorgeous and exquisite, which was a decoration for celebrating the birthday (Figure 1-5-2).

(三)《花卉册》/*Flower Album*

缂丝《花卉册》，十二开，每开长41cm，宽42cm，馆藏于故宫博物院。花卉册在本色地上缂织茶花水仙、玉兰海棠、牡丹、碧桃芝竹、百合剪春罗、荷花蜻蜓、踯躅、芝兰、秋海棠、剪秋纱、蝴蝶花、梅花等各色花卉，每开均缂织而成，无着笔处。画幅采用平缂、构缂、木梳戗、长短戗、凤尾戗和掼缂等缂织技法，缂工细致，润色自然，描画不求形似，具有简逸风格的文人花鸟画风（图1-5-3）。

Flower Album, as shown in Figure 1-5-3, had twelve sheets, each of which was 41 cm in length and 42 cm in width, which was collected in the Palace Museum. The album was woven with several flowers such as camellia, narcissus, magnolia, begonia, peony, peach blossom, bamboo, lilium, lychnis coronata thunb, iris japonica and plum blossom on the natural color ground, which these kossu processes were faultless and the images were vivid. These paintings adopt flat weaving, structure weaving, wooden-comb draw weaving, long-and-short draw weaving, phoenix tail draw weaving, Guan weaving and other techniques. They were a kind of literati's painting style of flowers and birds with a simple and elegant style.

图 1-5-3 缂丝《花卉册》(局部)
Kossu *Flower Album* (Partially)

（四）《仙山楼阁图轴》/Handscroll of Pavilions in the Immortal Mountains

缂丝《仙山楼阁图轴》长247cm，宽60cm。图在本色地上彩缂崇山、翔鹤、楼阁、人物、竹石、小草和碧波等纹样。采用平缂、构缂、搭缂、长短戗等技法缂织，配色适宜，缂织精细。图中山石的表现深受大青绿山水绘画的影响，勾勒无皴，以石青、石绿敷彩，其色彩厚重，格调富丽。画面又有云雾缭绕、祥鹤飞舞，仿若仙境（图1-5-4）。

Kossu *Handscroll of Pavilions in the Immortal Mountains* was 247 cm in length and 60 cm in width. Several patterns were painted on the natural color ground, such as mountains, flying cranes, pavilions, figures, bamboo, stones, grass and green waves. It was woven by the techniques of flat weaving, structure weaving, tapered weaving and long-and-short draw weaving, etc. The color matching was suitable and the weaving was exquisite. The rocks in the painting were deeply influenced by the landscape drawing, mainly in azurite and stone green, with heavy color and diversified styles. The painting was full of clouds, with several cranes flying, like a fairyland (Figure 1-5-4) .

（五）《赵昌花卉图卷》/ Zhao Chang's Flower Handscroll

缂丝《赵昌花卉图卷》长44cm，宽245cm。馆藏于故宫博物院。这件缂丝作品以赵昌之画为粉本，运用长短戗、木梳戗、凤尾戗、子母经和搭缂等技法缂织牡丹飞蝶、荷花、芙蓉翠鸟和鹊梅四组写生花鸟。设色清丽典雅，清劲秀逸，多采用三晕色的配色方法，如荷叶正面用墨绿，叶脉用草绿，背面用黄绿，色度

图1-5-4　缂丝《仙山楼阁图轴》
Kossu *Handscroll of Pavilions in the Immortal Mountains*

从深到浅逐层递减，过渡自然，表现出荷叶仰偃卷曲、柔媚秀雅的姿态和阴阳向背、光泽明暗的质感。飞蝶、栖鸟则捕捉其细部特征和灵动的神态予以精确的刻画，给人逼真酷肖的传神之感。这幅作品幅面宏大而手法细腻，画面无一处用笔，缂工极为精湛（图1-5-5）。

Kossu *Zhao Chang's, Flower Handscroll* was 44 cm in length and 245 cm in width, collected in the Palace Museum. This kossu artwork took Zhao Chang's painting as the manuscript, and wove four groups of flowers and birds, peony, flying butterfly, lotus flower, hibiscus, kingfisher and Sageretia by using the techniques of long-and-short draw weaving, wooden-comb draw weaving, phoenixtail draw weaving, Zi Mu warps weaving and tapered weaving techniques. The color was clear and elegant, matching by the method of three-colors bleeding. For example, the front of lotus leaf was dark green; the veins were grass green; the back was yellow-green. The color transition from deep to shallow was natural, showing the posture of the curling lotus leaf as well as its soft and beautiful posture. The textured lotus was produced. Flying butterflies and perching birds were captured their detailed features and smart demeanor. And then they were depicted carefully, presenting a vivid and cool feeling. This work was rich in its images and characterized by its delicate kossu skill, without any help of the painting brush（Figure 1-5-5）.

图1-5-5　缂丝《赵昌花卉图卷》
Kossu *Zhao Chang's Flower Handscroll*

◎ 思考题/**Questions for Discussion**

1. 明代缂丝创新了什么织造技法？/What innovations did the kossu techniques of the Ming Dynasty have?

2. 明代缂丝所用材料的特点有哪些？/What are the characteristics of the materials used for kossu in the Ming Dynasty?

3. 你对本节课所展示的明代作品印象最深刻的是哪幅？原因是什么？/Which of the Ming Dynasty works you were most impressed with in this lesson? And Why?

第六节　清代缂丝/Kossu Fabrics in the Qing Dynasty

清代缂丝

一、产生的背景/Background

清朝是满族建立的政权，但是统治阶级对汉文化有着很深的理解，积极继承和模仿明朝先进的文化。在工艺美术领域设立专门的机构，投入大量人力和物力模仿宋代作品，并大量研制新技术和新品种，在乾隆时期达到鼎盛。缂丝艺术在清代形成一次发展高潮，产量大幅增加，品种进一步丰富，缂织技术也有所创新。缂丝在清朝一度被皇家垄断御用，乾隆以后，缂丝面料的服饰在皇帝和后妃们的衣柜中大量出现，民间几乎不得见。在清朝宫廷内，缂丝常见于绘画、装饰品、服饰等。清代帝皇对书画艺术的喜爱直接推动了缂丝绘画在清代的全面发展，除了书画和宗教题材的缂丝织品，缂丝服饰、缂丝团扇等宫廷实用品也可谓是精美异常。

The Qing Dynasty represented a regime established by Manchu, one ethnic minority in China, but the ruling class had a deep understanding of Han culture. It was actively inherited and imitated the advanced culture of Ming Dynasty. During this time, specialized institutions were set up in the field of arts and crafts and large scale of manpower and resources were invested to copy works of the Song Dynasty, which led to the development of new skills and types. And it reached the peak in Qianlong period, with a substantial increase in output and types, promoting the innovation of kossu technology. Kossu fabrics could only be seen in the royal family in Qing Dynasty for a time. After Qianlong period, many costumes made of kossu fabrics appeared in the wardrobes of emperors and empresses, which were hardly seen among the ordinary people. In the palace of Qing Dynasty, kossu fabrics were common in paintings, decorations and costumes. The emperor's appreciation for painting and calligraphy in Qing Dynasty directly promoted the all-round development of kossu painting. Besides kossu fabrics themed by painting and calligraphy and religions, costumes, round fans and other daily use made by kossu were also exquisite.

清代生产帝后的袍服，纹样既保留明代风格，又掺入清代新风，具有过渡时期的特点。色彩上以黄蓝色系为主，对比强烈，表现出明显的清代风格。顺治三年清政府派人管理苏州、杭州的织造局。当时苏杭织造为顺治皇帝制作了大量服装，以至这些珍贵的缂丝龙袍还没有来得及穿用而一直存放于宫廷。

The costumes of emperors and empresses in Qing Dynasty not only retained the style of Ming Dynasty, but also applied the new style of Qing Dynasty, which had the characteristics of transitional period. The color was mainly yellow and blue with strong contrast, showing the style of Qing Dynasty. In the third year of Shunzhi, the Qing government sent people to manage the weaving bureaus in Suzhou and Hangzhou. At that time, the weaving bureaus made a lot of costumes for Shunzhi, but many of these precious dragon robes were stored in the palace before they could be worn.

康熙、雍正、乾隆时期政治稳定，经济繁荣，工艺美术领域也取得了很大的成就。乾隆时期缂丝艺术成就斐然。《石渠宝笈》是乾隆年间完成的中国书画著录。书中还收录以历代名家书画或以佛道题材的宗教书画为蓝本，用缂丝和刺绣工艺制作的优秀的织绣画。清代观赏性缂丝数量大增，尤其是乾隆时期摹缂乾隆皇帝的书法和绘画作品或为前人题字的作品占相当的比例。除了摹缂皇帝的作品外，这个时期释道人物、祝福祝寿等题材的缂丝作品也占据相当的比重。

During the Kangxi, Yongzheng and Qianlong periods, political stability and economic prosperity had been achieved, and great achievements had also been made in arts and crafts field. During the Qianlong period, the art of kossu had made outstanding achievements. *Shigu Treasure Collection* was famous descriptions of Chinese paintings and calligraphy during Qianlong period. They contained excellent works made by kossu and embroidery, which were based on famous paintings and calligraphy or Buddhist and Taoist themes. In the Qing Dynasty, the number of ornamental kossu works increased greatly, especially in the Qianlong period. The copies of emperor's calligraphy, paintings and inscriptions for last generations with kossu accounted for a considerable proportion. In addition to copying the emperor's works, the kossu works were themed by Buddhist and Taoist figures, and birthday celebrations also occupied a large proportion in this period.

受当时绘画、工艺美术以及西方装饰艺术的影响，缂织技术进一步丰富，艺人们又创造出缂丝加绣、缂金加绣、丝毛混织以及新型缂毛等新品种，使古老的缂丝艺术发扬光大。这个时期出现了"三蓝缂丝""水墨缂丝"等新的装饰品种。清代缂丝中心仍在苏州，集中在苏州的陆慕、蠡口一带，清宫所用的缂丝大多产于此。

Influenced by painting, arts and western decorations at that time, the kossu technology was further developed. There was appearing a lot of new types of kossu such as Embroidery with kossu, Embroidery with gold-kossu, silk woollen mixed knitting and new-type kossu wool, which helped to carry forward the ancient kossu technique. During this period, new techniques such as three-blue weaving and ink weaving appeared. In the Qing Dynasty, the kossu center was still in Suzhou, mainly in Lumu and Likou, where most of the kossu fabrics used in the palace were made.

清代后期由于国力衰微，对缂丝工艺品的需求不断减少，精工制作的缂丝织品更是凤毛麟角，随着缂丝商品化程度的提高这种倾向日益明显。为了节省时间和人力，又享受这种象征富贵豪华的织物，不得不想方设法减少复杂的制作工序，最直接的方法就是以画代织。这种倾向日益严重，到后期只织轮廓，花纹部分全部画出来，最终失去了缂丝原来的意义。

In the late Qing Dynasty, the demand of kossu fabrics was decreased gradually due to the decline of national power, and there were even few elaborate works. Influenced by the development of the commercialization of kossu, the production of kossu fabrics was fewer. In order to save time and manpower, people had to reduce complicated production processes to meet with their vanity. The most direct way was to replace weaving with painting, which was adopted more frequently later. And finally, only the outline was weaved and the pattern design relied on painting, which lost the original meaning of kossu.

二、产品特点/Characteristics of Products

（一）主要特征/Main Features

（1）清代缂丝技艺精细，运丝、用丝匀慎，紧密整齐，为前代所不及。

It was exquisite, well-proportioned, compact, regular, and superior to the former dynasties.

（2）色彩鲜艳、层次分明、沉而不浊、艳而不俗，阴阳、浓淡设色有别，富有层次的装饰效果和节奏美。

It was colorful, well-bedded, heavy but not vague and gorgeous but not vulgar. And it was also featured with reasonable accommodations in light and shade, and rich levels of decorations and rhythm beauty.

（3）取材极广，装饰手法多变，除山水、花鸟、人物和摹缂名人书画外，吉祥图饰更为盛行，并能吸收外来艺术因素和姊妹艺术之长，融为我用，独树一格。

It had a wide range of themes and changeable decorations. Apart from landscapes, flowers and birds, figures and famous paintings and calligraphy, auspicious decorations were more popular, and it selected the essence of foreign art and other similar art to make a unique style.

用缂丝加绣有两种不同工艺技法，如缂丝《三星图》就是富有地方特色和独树一格的作品，在发挥各自特色中协调统一。缂绘作品是大胆的改革和探索，使作品更完美、逼真，给人以美的感受，如1915年在巴拿马博览会展出的获奖作品缂丝《麻姑献寿图》，麻姑的面部开相和衣褶的明暗，运用笔墨晕染。缂画就是缂出物像的大概轮廓，余部均用笔描绘或以彩敷之，此法在宋代早已用之。

There were two different types of embroidery with kossu, such as *Three Stars*, which was works with local characteristics and unique style. The kossu painting was the result of bold reforms and explorations, which made the works more perfect, realistic and beautiful. For example, the winning work *Ma Gu's Celebration of Birthday*, exhibited at Panama Expo in 1915, depicted Ma Gu's face and the light and shade of clothes pleats with painting. Kossu painting referred to delineating the outline of the object and finishing the rest part with ink or pigment, which had already been applied in the Song Dynasty.

（4）清初缂丝在继承宋、明优良传统的基础上，更日趋精进，品种、技艺、造型、纹饰及民族传统风格，图案与写实结合，均发展较快。

On the basis of Song and Ming Dynasties, the kossu fabrics in the early Qing Dynasty got further refined and developed in many aspects, such as types, skills, shapes, ornamentation, traditional nation styles, combinations of pictures and realism.

（二）技法工艺/Techniques

缂丝艺术在清代形成一次发展高潮，产量大幅增加，品种进一步丰富，缂织技术也有所创新。清代缂丝技术有很大的发展，其中之一就是创造了双面"透缂"技术，这种缂丝工艺制作出来的织物两面花纹是相同的，线条清晰平整，比较适合装饰插屏和扇子，所以清代有大量的团扇都使用了"透缂"技术。缂丝团扇立体感强，加上题材都是人们喜闻乐见的，所以其艺术和观赏价值完全可以和缂丝绘画等分庭抗礼。清代宫廷缂丝继承了明代的缂丝技法，并在其基础上不断发展和创新，清代的缂制精密牢固，技术精进，用丝细匀，题材内容丰富，设计繁缛，逐渐成为融缂丝、刺绣、绘画为一体的综合性艺术，色彩艳丽，更富变化，堪称中国古代缂丝作品中的精品。清代的缂丝，在技法、表现手法、用色以及艺术形式上有以下几个特征：

Kossu techniques formed a development climax in the Qing Dynasty, with a substantial increase in output, further enrichment of types and innovation in techniques. One of the development was the creation of double-sided transparent kossu. This kind of fabric made by kossu had the same pattern on both sides with clear and flat lines, which was more suitable for the production of decorative screens and fans. Therefore, a large number of round fans in the Qing Dynasty were applied transparent kossu technology. The round fans made by kossu were vivid and the themes of them were accepted and loved by ordinary people, so the artistic and ornamental value could completely compete with kossu painting. The palace kossu fabrics in Qing Dynasty inherited the technique in Ming Dynasty, and made it continuously developed and innovated. The kossu fabrics in Qing Dynasty were exquisite, well-designed and various in themes. It had gradually become a comprehensive art integrating kossu fabric, embroidery and painting, which was colorful and changeable, and could be called high-quality works of ancient Chinese kossu fabric. The kossu fabric in Qing Dynasty had the following characteristics in technique, expression, color and art:

（1）双面"透缂"技术得到进一步发展。清代的缂丝品由于广泛地向生活实用方面拓展，如服饰、扇子、屏风等，在艺术效果上更追求双面花纹清晰、外观平整，因此，两面花纹相同的"绣缂"技术较之前代更为娴熟。

The double-sided transparent kossu fabrics had been further developed. In the Qing Dynasty, the productions were widely expanded to practical aspects of life, such as costumes, fans, folding screens, which pursued clear double-sided patterns and smooth appearance in artistic effect. Therefore, the embroidery with kossu fabric which had the same double-sided patterns was more skilled than the previous generation.

（2）为了增强色彩效果和体现色彩的变化，大量采用两种不同深浅或不同颜色的色丝合捻而成的"合色线"，以表现花纹图案的质感和明暗变化。

In order to enhance the color effect and reflect the color change, a large number of "colored lines" formed by twisting two kinds of color wires with different depths or colors were used to show the texture and light and dark changes of patterns.

（3）缂、绣、绘不同表现手法的结合运用更为广泛。尤其是清中晚期的缂丝品，缂与绣、绘并用的表现手法运用广泛。不仅欣赏性作品大量加墨敷彩渲染，甚至服饰和其他实用缂丝品也用笔墨晕染，有的作品中的人物，缂匠只缂织外轮廓线，其余全用笔墨描绘。

The combination of kossu fabric, embroidery and painting was more widely used, especially in the middle and late Qing Dynasty. At this time, ink and pigment were applied to the kossu fabric, including ornamental works, costumes and other practical products. Some figures were only drew the outline and the rest parts were finished by ink and pigment.

（4）清中期出现的"三色金"缂丝，则是指缂丝色线用材方面的特色。"三色金"缂丝指采用赤圆金、淡圆金及银丝为纬线缂织的作品，一般采用深色地子，花纹图案以金银线缂织，作品光灿夺目，价值不赀。晚清出现了"三蓝缂丝"和"水墨缂丝"。"三蓝缂丝"指在素地上用深蓝、品蓝、月白（浅蓝）三色退晕的戗缂方法缂成各种花纹，并用白色勾边。"水墨缂丝"指在浅色素底上用黑、深灰、浅灰三色退晕的戗缂法织成各种花纹，并用深色或白色勾边。以上两色，成为晚清缂丝作品的流行色。

The three-color gold kossu, appeared in the middle of Qing Dynasty, was a kind of kossu fabric applying three colors: red round gold, light round gold and silver. And it generally chose dark color, design graph and metallic fabric, resulting in a dazzling and valuable work. In the late Qing Dynasty, three-blue kossu and ink kossu appeared. three-blue kossu referred to the method of weaving various patterns on plain cloth with dark blue, royal blue and light blue, and delineated the outline with white. Ink kossu referred to weaving various patterns on a light base with black, dark gray and light gray, and delineated the outline with dark or white. The above two types of kossu fabric were the most popular in the late Qing Dynasty.

（三）纹样/Patterns

清代制作了大量供皇室穿着的缂丝服饰及供皇家供奉的缂丝唐卡，均是缂丝中的珍品。缂丝作品的题材趋向临摹宋、元、明各朝名画和仿制缂丝名作，形式上出现大型巨幅壁挂、中堂和精巧的小品，技法更趋成熟。

In the Qing Dynasty, the large production of kossu costumes and kossu Thang-ga for the royal family were made, which were treasures of kossu. The development of kossu tended to copy famous paintings of the Song, Yuan and Ming Dynasties and the former excellent works. And the appearance of huge wall hangings, middle-sized and exquisite sketches showed the gradually matured techniques.

随着宗教日益世俗化，佛教、道教题材的缂丝作品大幅增加。实用品的种类更加丰富，宫廷服装和官服大量使用缂丝，并使用金线、银线以及孔雀羽线织造。宗教性缂丝作品经常与文字相配合，同时为了细致地描绘细部，有时会采用加绣和敷彩、敷金等方式来描绘细部，使局部更加形象、生动。书画型缂丝作品在清代达到极度繁荣的高峰，可分为书画相结合的缂丝作品、纯书法缂丝作品、纯绘画缂丝作品，传世的清代作品有《贵妃醉酒》《牛郎织女》等。

With a closer connection between religion and life, the kossu works featured with Buddhist and Taoist had greatly increased. There were more kinds of practical products, and many court clothes and official clothes were made by gold thread, silver thread and peacock feather thread. Religious kossu fabric was often combined with writing. At the same time, embroidery, pigment and gold were often applied to describe the detail parts, which could make a more vivid and lively work. The development of kossu fabric combined with writing and painting reached a peak in the Qing Dynasty. It could be divided into three types: kossu fabric with writing, kossu fabric with painting, the combination of the two. The works handed down at that time included *Drunkened Concubine*, *the cowherd and the weaving maid* and so on.

三、名家名作/Masterpieces

王新亭，清同治、光绪年间吴县（今苏州吴中区）缂丝名匠，陆墓张花村人，一度为宫廷匠师，曾缂织《八仙庆寿》等服饰。

Wang Xinting was a famous craftsman from Zhanghua Village in Lumu, Wu County （Wuzhong District, Suzhou）during Tongzhi and Guangxu period (in the Qing Dynasty). He once served for the palace and made clothes with kossu, such as *Eight immortals to Celebrate Birthday*.

王锦亭，王新亭之子，晚清至民国初年缂丝名匠，兼长丹青。早年曾制作清廷御用缂丝品，代表作《麻姑献寿图》参展1915年巴拿马国际博览会并获奖。

Wang Jinting, the son of Wang Xinting, was also famous in the late Qing Dynasty to the early Republic of China, who was also a famous painter. In his early years, he made kossu fabrics for the Qing palace, and his representative work *Ma Gu's Celebration of Birthday* participated in Panama International Expo in 1915 and won an award.

《麻姑献寿图》绘身着红披风的麻姑，古朴静穆、仪态端庄，健康清新，令人瞩目。图中设色艳丽华贵，富有装饰性，线条顿挫刚劲，富有表现力，与所绘人物相得益彰。民间为妇女祝寿时常常绘《麻姑献寿》的图画相赠以示祝福（图1-6-1）。

Ma Gu's celebration of Birthday depicted a lady named Ma Gu who dressed a red cloak and looked very beautiful and healthy with great elegance. The colors in the picture were gorgeous with decorative value and the lines were bold with expressive power, and all of them looked harmonious. When celebrating women's birthday, people often drew the *Ma Gu's celebration of Birthday* to express their blessings（Figure 1-6-1）.

缂丝《芦雁图》长104cm，宽45cm。本色地上彩缂芦雁和小草等纹样。画面基本以绿色、棕色、湖色、香色、墨绿等稳重的色丝搭配，构图简洁，造型写实。作品采用平缂、构缂和搭缂等技法缂织，作者娴熟地处理线条的弯转变化，将被风吹拂下的芦苇的飘动和叶子翻折下的明暗变化生动地表现出来，颇富天然之趣。两只芦雁神态安详，怡然自得。画面动静结合，充满和谐自然的逸趣（图1-6-2）。

Kossu *Wild Goose in the Reeds,* was 104 cm in length and 45 cm in width. The pattern design such as wild goose, and grass were made by colored thread. The picture was basically adopted stable colors such as green, brown, light green, dark brown and dark green, presenting a concise composition and realistic style. Techniques such as flat weaving, structure weaving and tapered weaving were applied in it. The author skillfully handled the bending changes of lines and vividly showed the fluttering of reeds under the wind and the light and shade changed under the folding of leaves in a natural way. Two calm and contented Lu Yan were described in the picture. The combination of activity and stillness made the picture look harmonious and full of natural interest（Figure 1-6-2）.

图1-6-1 缂丝《麻姑献寿图》
Kossu *Ma Gu's Celebration of Birthday*

缂丝《仇英后赤壁赋图卷》长30cm，宽498cm。此图卷由缂丝艺人以仇英《后赤壁赋》画为蓝本缂织而成，详尽地描绘出苏轼在初冬时节夜游赤壁的场景。画面共分8段，前三段表现苏轼重游旧地的欢快，后五段则借景写情，寄人生如梦之感慨。图卷采用平缂、长短戗、构缂、搭梭、掼等多种缂织方法，使用了20多种颜色的丝线，所织人物形象生动，景物写实，山石勾中有皴，以石青、石绿等色为主，表现出仇英的小青绿山水风格，反映了清代缂丝艺人高超的工艺水平（图1-6-3）。

Kossu *Qiu Ying Post-Red Cliff Ode* was 30 cm in length and 498 cm in width. This scroll was made by kossu based on Qiu Ying's painting *Post-Red Cliff Ode*, which elaborately depicted Su Shi's night tour of Red Cliff in early winter. The picture was divided into eight parts. The first three parts showed Su Shi's pleasure of revisiting the old place, while the last five parts took advantage of the scenes to express that life was like a dream. The scroll was rich in kossu skills, such as flat weaving, long-and-short

图1-6-2 缂丝《芦雁图》
Kossu *Wild Goose in the Reeds*

draw weaving, structure weaving, shuttling weaving, Guan weaving and applied threads more than 20 colors. The finished work presented a realistic style in the scenery with vivid figures. And the colors texture of the rocks adopted were mainly azurite, malachite and brown ochre, reflecting the fresh and clean style of Qiu Ying's painting and the excellent skills in the Qing Dynasty (Figure 1–6–3).

图1-6-3　缂丝《仇英后赤壁赋图卷》（局部）
Kossu *Qiu Ying Post-Red Cliff Ode* (Partially)

缂丝《秋桃绶带图》长198cm，宽59cm。图以桃和绶带鸟为题材缂织，寓祝寿之意。作品原稿具有明代吴派画家陆治的"淡彩写生"的风格，工笔勾勒与设色没骨法相兼。艳丽的花卉、桃实与淡彩晕染的坡石、溪流相映成趣，绶带鸟笔墨工丽、写实，配色文秀雅致，属于文人画中工笔淡彩的画风，反映出作者娴静高雅的性情和蕴藉风流的艺术追求。为追摹原画稿笔意，此作缂技灵巧多变，采用平缂、搭缂、长短戗、掼缂和木梳戗等多种技法，使色彩晕染更趋自然。

图1-6-4　缂丝《秋桃绶带图》
Kossu *Autumn Peaches and Chinese Paradise-flycatchers*

尤其是在缂织桃实尖和晕染山石时，采用华丽精美的凤尾戗技法，从而更具装饰性。缂织细微处，如海棠花残叶，则频繁而灵活地变换色丝小梭，以求写实的效果。花朵和叶脉则用构缂勾勒以表现工笔画之趣味（图1-6-4）。

Kossu *Autumn Peaches and Chinese Paradise-flycatchers* was 198 cm in length and 59 cm in width. It was themed by peaches and paradise-flycatchers, which meant celebrating the elder's birthday. The original manuscript had the style of light color to paint by Lu Zhi, a painter from Wu Painting School in the Ming Dynasty and it was skilled in the combination of fine brushwork and Mogu Method the method of painting with color but without brush. Bright colored flowers, peaches, light-colored slopes and streams in the picture contrasted finely with each other. Paradise-flycatchers were beautiful and vivid with light color, which belonged to the style of combination with fine brushwork and light color, reflecting the author's quiet and elegant temperament and romantic artistic pursuit. In order to imitate the charming of the original painting, the work was finished with dexterous and changeable skill, and various techniques such as flat weaving, tapered weaving, long-and-short draw weaving, Guan weaving and wooden-comb draw weaving were adopted to make the color mix more natural. Especially when weaving peach tips and rocks, the gorgeous and exquisite phoenix tail weaving technique was adopted, which was more decorative. The subtle parts of the tapestry, such as the residual leaves of Begonia flowers, changed the silk shuttle frequently and flexibly in order to achieve a realistic effect. The flowers and veins of leaves were outlined with structure weaving to show the interest of meticulous painting（Figure 1-6-4）.

《鸡雏待饲图》馆藏于故宫博物院，长64cm，宽36cm。此图为一件集书法、绘画、缂丝、缂毛等艺术手段的合璧之作。上部为乾隆御笔题跋，下部为仿南宋李迪所绘《鸡雏待饲图》。书与画相得益彰。此图在缂织技法上追摹李迪原作写实的神韵，着意表现鸡雏渴望、无奈、无助的神态，并采用丝毛合捻之线，施以长短戗、平缂、搭梭等技法晕色，使鸡雏羽毛的层次和质感表现得尤为真实。摹缂御笔文字则用丝线，虽然仅用平缂、搭缂技法，但缂工细致传神（图1-6-5）。

图1-6-5　缂丝《鸡雏待饲图》
Kossu *Chicks to Be Fed*

Chicks to Be ed was collected in the Palace Museum, with a length of 64 cm and a width of 36 cm. This picture was a combination of calligraphy, painting, kossu fabric and kossu wool. The upper part was the inscription of Qianlong, and the lower part was the copy of *Chicks to Be Fed* painted by Li Di in the Southern Song Dynasty. And the two parts matched very well. This picture followed the realistic style of the original work with kossu techniques, and emphasized the eagerness and helpless ness of chickens. It also adopted the mixed thread of silk and wool, painting with specialized techniques such as long-and-short draw weaving, flat weaving and shuttling weaving to make the chickens' feathers more vivid. Silk thread was used to copy the writings of the emperor. The techniques applied were only flat weaving and tapered weaving but very exquisite and lively（Figure 1-6-5）.

◎ 思考题/Questions for Discussion

1. 清代缂丝绘画在清代全面发展的直接原因是什么？/What is the direct reason for the

all-round development of kossu painting in the Qing Dynasty?

2. 清代缂丝产品的特点有哪些？/What are the characteristics of kossu works in the Qing Dynasty?

3. 清代缂丝在技法、表现手法、用色等方面有何特征？/What are the characteristics of kossu techniques in the Qing Dynasty in skill, expression and color?

第七节　近现代缂丝/Kossu Fabrics in Modern Times

一、产生的背景/Background

近现代缂丝

辛亥革命推翻帝制，服装改朝换代加上抗日战争爆发等原因，缂丝品市场越来越小，甚至渐趋没落，能织造者不足六十人，技艺精湛者二三十人而已，千年缂丝绝技濒临消亡。

Due to the Revolution of 1911, the change of clothing and the outbreak of anti-Japanese war, the market of kossu fabrics were getting smaller and even declining. There were less than 60 people who could make kossu fabrics, and only 20 or 30 people with exquisite skills. The unique skill of kossu in the millennium was on the verge of extinction.

中华人民共和国成立后，缂丝工艺重获新生。在国家和地方政府"保护、提高、发展"的方针指引下，苏州缂丝枯木逢春。苏州缂丝艺人们通过积极发掘、抢救、传承优秀缂丝工艺，使固有的传统缂丝技艺得以继承和发展。苏州市文联邀请专家顾公硕、贺野、高伯瑜等人组成调查组，分赴郊区产地寻访老艺人，对缂丝工艺进行了抢救性保护工作。1954年成立了"苏州市文联刺绣小组"（即苏州刺绣研究所前身），邀请了缂丝制作经验丰富、技艺水平高超的老艺人王茂仙、沈金水等缂丝名匠在拙政园内当众表演缂丝技艺，得到社会重视。1955年5月起，由王茂仙带领从陆慕、蠡口招收的12名缂丝艺人，与刺绣组合成立生产组。同年2月苏州市文联在市刺绣合作社内组织缂丝生产，召集缂丝艺人加入市刺绣合作社，不久生产组扩大为苏州刺绣工艺美术生产合作社，并由工艺美术研究室领导，研究技术创新，培养新生力量。1956年，在民间发展了一批缂丝人员，开始带徒授艺，培养新一代缂丝人才，现代缂丝艺人王金山就是沈金水当时的徒弟之一，共有20余人，20多台缂机。先后缂织了《玉兰黄鹂》《牡丹双鸽》《博古》《双鹅梅竹》等一批缂丝艺术品，现藏于南京博物馆。20世纪70年代，苏州先后建立了苏州缂丝厂、吴县东山缂丝厂、蠡口缂丝厂、黄桥缂丝厂、陆慕缂丝厂五大缂丝龙头厂，共有600余人，缂机600多台，还注册了金兰、和合、天宫等商标。

However, the founding of the People's Republic of China brought about a new turn for the development of kossu technique and revitalized the kossu market. Kossu industry in Suzhou started to find a new way under the guidance of the policy of "protecting, improving and developing". By craftsmen's efforts on exploring, inheriting and renovating excellent kossu skills, the traditional kossu technique could be developed and inherited. Literary Federation of Suzhou invited experts such as Gu Gongshuo, He Ye, Gao Boyu to set up an investigation team, and went to the suburban to visit old craftsmen, aiming to protect the kossu techniques. Embroidery Group of Suzhou Literary Federation (the predecessor of Suzhou Embroidery Research Institute) was established in 1954. The experienced craftsmen Wang Maoxian and Shen Jinshui were invited to show their excellent kossu skills in public in Humble Administrator Garden, which gained social attention. Since May 1955, Wang Maoxian had led 12 kossu workers employed from Lumu and Likou to set up a production group on the basis of embroidery group. In February, the Literary Federation organized people to produce kossu fabrics in the Embroidery Cooperative to call kossu workers to join the Cooperative. Soon, the production group expanded to Suzhou Embroidery Arts and Crafts Production Cooperative Association, which was led by the Arts and Crafts Research Office to study technological innovation and cultivate new forces. In 1956, a group of kossu craftsmen were developed and they began to teach others to train a new generation of kossu talents. Wang Jinshan, modern kossu craftsman, was one of Shen Jinshui's students at that time. With more than 20 workers and 20 machines, a number of kossu artworks such as *Magnolia and Loriots, Peong and Twin Pigeons Bogu, Double Geese with Plums and Bamboos* had been finished successively, which are now in Nanjing Museum. In the 1970s, five kossu factories were established one after another in Suzhou, Wu County Dongshan, Likou, Huangqiao and Lumu, which were named the five leading kossu factories with more than 600 people and 600 machines. They also registered trademarks such as Jinlan, Hehe and Tiangong.

20世纪60年代，苏州缂丝艺术有了进一步的发展，党和政府陆续将分散在各地的手工作坊以及手艺人集中起来，根据工艺门类组织了缂丝、刺绣等多个生产合作。同一时期，苏州的手工艺行业，从生产适应市场需求的小商品逐步发展为工艺品出口，企业数量增多，至20世纪60年代中期已经具有相当规模，并且在苏州郊区有一支庞大的缂丝外加工队伍。为了保护织造精良的宋代缂丝作品，1963年至1965年，苏州缂丝艺人王金山、李荣根、陶佳燕等人到故宫博物院对部分宋代经典缂丝作品进行了充分的研究，并复制成功，不仅积累了宝贵的经验，还对研究宋代缂丝提供了技术性的资料，尝试摹缂古代绘画作品。苏州缂丝研究所的沈金山等人缂织成功宋徽宗的绘画《柳鸭芦雁图卷》，使现代缂丝艺术迈出了一大步。

In the 1960s, kossu techniques in Suzhou made a further development. The Communist Party and government gathered manual workshops and craftsmen scattered in various places to establish production cooperatives such as kossu cooperative and embroidery cooperative according to the different techniques. During the same period, Suzhou's handicraft industry gradually developed from the production to meet the market demand to the export of handicrafts, and the number of enterprises also increased. By the mid–1960s, it had reached a considerable scale. What's more, there was a huge kossu subcontract team in the suburbs of Suzhou. In order to protect exquisite kossu fabrics of the Song Dynasty, some famous kossu craftsmen such as Wang Jinshan, Li Ronggen and Tao Jiayan went to the Palace Museum to fully research kossu works in Southern Song Dynasty, and then copied these works, which not only accumulated valuable experience for the development of kossu technique, but also provided technical information for studying the kossu technique in Song Dynasty and helped to copy ancient paintings with kossu. Shen Jinshan from Suzhou Kossu Fabrics Research Institute led his team to copy the emperor Song Huizong's painting *Willow and Geese in the Reed* successfully, which promoted the development of modern kossu greatly.

1984年，苏州六位缂丝艺人历时三年复制成功《十二章福寿如意衮服》，现珍藏于定陵博物馆。

In 1984, six kossu craftsmen in Suzhou spent 3 years finishing the copy of *Ceremonial Robe of Rulers with Twelve Patterns Good Fortune*, which is now collected in Dingling Museum.

20世纪70年代末至20世纪90年代初，是苏州缂丝艺术发展的辉煌时期。20世纪80年代，随着我国对外开放的不断深入，工艺美术的外贸出口任务猛增，高档日用品、和服腰带由于受到日本客户的青睐，出口需求量逐年上升，各缂丝厂家抓住这一契机，为确保外贸出口任务的完成，纷纷增加工人和添置缂机设备。当时，苏州缂丝人员达1万之多，超过历史上任何朝代，缂丝生产空前繁荣，几乎形成"村村有工厂，家家有机台"的规模。

From the late 1970s to the early 1990s, it was a brilliant period for the development of Suzhou kossu technique. In the 1980s, with the deepening of China's opening to the outside world, the export of arts and crafts increased sharply. Because of the favor of Japanese customers, the export demand of high–end daily use and kimono belts increased year by year. All kossu manufacturers took active actions to seize the opportunity such as employing workers and buying new equipment to ensure the export. At that time, there were the greatest number of kossu workers to more than 10,000 in Suzhou in history. The production of kossu fabrics was up to a golden time unprecedentedly, almost "every village had a factory and every family had a kossu machine".

进入20世纪90年代，由于缂丝业的盲目发展、个体户的无序竞争，粗制滥造的缂丝品充斥市场，加上受世界尤其是亚太地区经济危机的影响，工艺美术品外贸任务日趋下降，生产萎缩，缂丝技艺人员外流，绝大部分企业停业转向，艺人转行转业，制作艺术精品的艺人越来越少，缂丝古艺，再一次由盛转衰，技艺的传承，成为缂丝业亟须解决的头等大事。

In the 1990s, due to the blind development of kossu industry, disorderly competition of individual businesses, the market was flooded with shoddy kossu products. The kossu industry had gone down. In addition, the outbreak of economic crisis in the world, especially in Asia–Pacific region, export and production of arts and crafts decreased day by day, caused an outflow of skilled workers, even some kossu store had to be closed or turn to another industry. Gradually, there were fewer and fewer artists making fine arts. The ancient art of kossu had once again changed from flourish to decadence, and the inheritance of skills had become the top priority that needed to be solved urgently in the kossu industry.

1996年，南通工艺美术研究所王玉祥自己出资成立了缂丝工作坊，用现实的经济市场来滋养千年中华绝技。2004年，王玉祥的缂丝工作坊与日本京都西阵织株式会馆成为友好合作机构。2007年，成立专门的缂丝研究机构，命名为"宣和缂丝研制所"。2008年，宣和缂丝研制所织造的缂丝品登上了中国国际时装周。2009年，宣和缂丝研制所以团体研究机构加入联合国教科文民间艺术国际组织。宣和缂丝研制所与王玉祥用一种崭新的方式延续着缂丝这一古老的技艺。

In 1996, a kossu workshop, funded by Wang Yuxiang of Nantong Arts and Crafts Research Institute, was established to develop the Millennium unique skill—kossu skill. In 2004, the workshop became a friendly cooperation partner of the Nishijin Textile Center in Kyoto, Japan. In 2007, a specialized kossu research institute was set up, named "Xuanhe Kossu Research Institute". In 2008, the kossu works made by Xuanhe Kossu Research Institute were listed in the China Fashion Week. In 2009, it joined IOV (international organisation rganisation für volkskunst) as a research group. Thus, Xuanhe Kossu Research Institute and Wang Yuxiang developed the ancient technique of kossu fabrics in a new way.

二、产品特点 /Characteristics of Products

随着社会的不断进步、缂丝技艺的不断提高，现代苏州缂丝在原料和工具上有了改进，在技艺上有了创新，色彩和图案题材上也有所变化，形式也变得更为丰富。

With the advancement of society and kossu technology, Suzhou kossu in modern times has improved a lot in many aspects, such as materials, tools, skills, colors and pattern themes.

（一）材料 /Materials

传统缂丝上所用的经线，基本是无须脱胶和练熟的生丝，纬线是需要脱胶和练熟再染成各种彩色的丝线。随着工艺的不断发展，出现练熟的熟经线，也就是熟经熟纬。由于采用的经线原料和质地不同，缂丝作品手感也不相同。用无须脱胶的生丝做经线，质地比较硬，完成后的作品较为挺括；用脱胶的熟经线缂织出的作品，质地较为柔软。艺人王金山提到，现在熟经熟纬的方法主要运用在服装、围巾、披肩、挂帘等要求质地比较柔软的产品上。

The warp adopted in traditional kossu technology was mainly silk thread without degumming and refining but the weft is needed and then dyed into various colored silk threads. With the development of technology, both refined warp and weft were adopted in the production. Due to the different materials and textures of warp, the softness of kossu fabrics was also different. The kossu fabrics which applied warp without degumming were relatively hard in texture and the finished products were usually neat and crisp, while the fabric which applied refined warp with degumming was often soft in texture. Wang Jinshan said that refined warp and weft was mainly used in clothing, scarves, shawls, hanging curtains and other soft-needed products.

苏州缂丝在材料上增加了羊绒，和丝线相比羊绒的颜色比较重，黑色显得毛茸茸的，适用于表现特殊效果。在金地上，用黑色羊绒显得更厚重、浓烈，一般丝线达不到这种要求，因此材料的选择根据作品需要来决定。

Now cashmere appeared as a new material in Suzhou kossu. Compared with silk thread, cashmere was darker and looked fluffier, which was often used to show special effects. The adoption of black cashmere on the gold tissue made the work look heavy with good texture which the silk thread could not reach. Generally speaking, the choice of materials depended on the needs of the works.

（二）技法工艺/Techniques

现代缂丝艺人继承宋代的传统技法，不用补笔，全部使用彩纬织造，与明清以后大量的补笔作品形成鲜明对比。

Modern kossu craftsmen inherited the traditional techniques of the Song Dynasty and the threads used were all colored weft without supplementary technique. It formed a sharp contrast with the works which applied supplementary technique in great quantity after the Ming and Qing Dynasties.

在传统技法基础上，艺人们通过反复实践开创了新的技法，使古老的缂丝艺术发扬光大。1965年《鲁迅像》《白求恩大夫》等作品采用长短戗和抽丝并色等技法，充分表现了人物面部神态。1973年开始，缂丝组俞家荣、李壮飞与苏州市缂丝研究所王俊儒等协作，攻克曲纬、整边等技术难关，到1974年试制成功素地电动缂丝机，由此成立了机缂组，生产素地缂丝腰带，产品外销供不应求。

Based on the traditional techniques, craftsmen hed developed new techniques through repeated practice to carry forward the ancient kossu technology. In 1965, *the portrait of Lu Xun and Dr. Bethune* adopted techniques such as long-and-short draw weaving, spinning and shade pitching to express the facial expression of characters, and then gained successful experience.

Since 1973, Yu Jiarong and Li Zhuangfei of the kossu group had cooperated with Wang Junru of Suzhou Kossu Research Institute to overcome technical difficulties such as weft bending and edge trimming. By 1974, the plain ground electric kossu machine had come out after repeated trials. Therefore, the machine kossu group was established to make kossu belts on plain tissues and the export is always in short supply.

1982年，苏州缂丝厂工艺师王金山开创了异纬织造法，缂制成功双面三异作品《牡丹—茶花—双蝶》，如图1-7-1所示，这幅作品收藏在中国工艺美术馆。作品在金地上反映三个截然不同的画面。第一个画面是花，一面是牡丹花，另一面是山茶花；第二个画面是蝴蝶，一面是橘黄色无尾巴的蝴蝶，另一面是黄色带尾巴的蝴蝶；第三个画面是印章，一面是"缂丝"字样，另一面是制作者姓名"王金山"。为了织造这一新品种，首先对传统缂丝机进行了局部改造，同时发展了缂丝多种新技法，如异纬法，以适应织造变化的需要。

In 1982, Wang Jinshan developed the different weft kossu weaving technique and successfully completed the work characterized by double-sided patterns, *Peony-Camellia-Double Butterflies*, as shown in Figure 1-7-1, collected in the China Arts and Crafts Museum. In this work, three different patterns were designed on the gold tissue. The first pattern was a flower with peony as one side and camellia as the other side. The second pattern was a butterfly with an orange butterfly without a tail as one side and a yellow butterfly with a tail as the other side. The third pattern was a stamp with the word "kossu" (in Chinese character) as one side and "Wang Jinshan" (in Chinese character) as the other side. In order to develop this new type of kossu fabrics, the traditional kossu machine was transformed partially and many new kossu skills were developed at the same time, such as different weft to meet the needs of productions.

（a）山茶花　　　　　　　　　　　（b）牡丹花
Camellia　　　　　　　　　　　　Peony

图1-7-1　缂丝《牡丹—茶花—双蝶》
Kossu *Peony-Camellia-Double Butterflies*

1984年，艺人们经过悉心研究，又成功织出两面图案完全不同的全异《寿星图》，即两面地色、图案、印章都不同，如图1-7-2所示。一面以清代画家任伯年的作品为蓝本，银红色地，织一手捧寿桃的寿星，并有任伯年的印章；另一面以吴昌硕的作品为蓝本，金色为底色，织黑色篆体字"寿"字，并有吴昌硕的印章。技术上使用多色合纬法，使两面纹样框架不受经纬规则的限制，小面积任意变化。这种新工艺的成功是全异作品成功的技术关键，为今后开创缂丝新品种、新花色开拓了新路子。

In 1984, *Longevity* with completely different patterns on both sides, was finished successfully after the full research by the craftsmen. The two sides were different in ground color, pattern and stamp, as shown in Figure 1–7–2. One side was based on the works of Ren Bonian, a painter in the Qing Dynasty and applied silver red ground to depict a god of longevity with a peach in his hand and then sealed by Ren Bonian. The other side was based on Wu Changshuo's calligraphy and chose gold as the ground color with the black seal script "Shou" (in Chinese character) on it and then sealed by Wu Changshuo. The method of weft mixed with more than two colors was used to reach that the two-sided pattern frame was not limited by the length and width and then could change freely in a certain scope. The development of this new skill was necessary to the works with different patterns, which committed to enriching the diversity of types and colors of kossu fabrics in the future.

（a）寿字　　　　　　　　　　　　　　　　（b）寿星
The Character for "Longevity"　　　　　　　　The God of Longevity

图1-7-2　缂丝《寿星图》
Kossu *Longevity*

（三）纹样/Patterns

随着苏州缂丝技艺的不断发展，缂丝的图案题材也从程式化的实用图案、山水花鸟逐步拓展为抽象风景等图案，并在艺人的努力下，不断扩宽了缂丝的题材。

With the development of Suzhou kossu technology, the pattern theme of kossu work had gradually expanded from practical patterns, landscape, flowers and birds to abstract landscape, and had constantly enlarged the pattern types.

20 世纪 90 年代，吴县缂丝总厂老艺人徐祥山、青年艺人王嘉良和马惠娟以及设计师马超骏对传统缂丝题材进行大胆创新，成功将国画大写意题材缂丝《乌龙取水图》、泼墨图题材《泼墨乌龙》以及披毛动物题材《虎啸图》缂织成功。虎是缂丝行业从未触及过的披毛动物题材，马惠娟借鉴了刺绣技艺中散套针法并运用到缂丝传统的长短戗、斜戗之中，使纬线呈放射状伸展，增强了虎毛的质感，解决了缂织披毛动物的一大难题，扩宽了缂丝的题材。

In the 1990s, guided by the experienced craftsman Xu Xiangshan, young craftsman Wang Jialiang and Ma Huijuanan and designer Ma Chaojun, the reform and innovation of traditional pattern themes was carried out. A series of works were reproduced successfully, such as the work of freehand brushwork in traditional Chinese painting *Dragon Drawong Water*, the work of splash-ink painting *Splashed-ink Dragon*, and the work themed by hairy animals *Roaring Tiger*. Tiger had never appeared as a pattern theme in kossu work. Ma Huijuan learned the skill of loose needing from embroidery technology and applied it to the traditional skills of long-and-short draw weaving and oblique weaving, which made the weft extending radially and enhanced the texture of tiger hair, solving a major problem caused by animals with long hair and diversifying the themes of kossu products.

缂丝人物题材，过去以宗教、神仙以及祝寿比较多见，也有山水人物，但从未缂织过人物肖像。20 世纪 90 年代，马惠娟以摄影照片为底稿，将《日本东大教授的人像》在木制缂机上缂织成功，是国内第一幅以人像为主题的缂丝精品，随后又缂织了《东洲人像》，拓宽了缂丝的题材。

In the past, the themes of figures were more common in religion, immortals and birthday celebrations. There were also landscape figures, but the portrait had never applied as a pattern theme. In the 1990s, Ma Huijuan successfully finished the kossu work named *Portrait of Professors in Tokyo University* with a wooden kossu machine based on a photo. It was the first exquisite kossu work with the theme of portrait in China. After that, she completed the work—*a portrait of Dongzhou.* These kossu works themed by portraits undoubtedly enlarged the diversity of pattern themes.

传统图案中有题诗、题词，并有印鉴。印鉴多种多样，有原书画作者印鉴、有诗词作者印鉴、题款者印鉴、鉴赏者印鉴等，随着现代缂丝艺人地位的提高，在现代缂丝作品中，也增加了缂织者姓名印鉴或者缂织者工作室印鉴，还可根据个人要求加入收藏者印鉴。

There were poems and inscriptions with stamps in the traditional patterns. And these stamps were often from various people including the author of the original calligraphy and painting, the author of the poem, the inscription author and the appreciator. With the improvement of the status of modern kossu craftsmen, the stamps of craftsmen or the kossu studio also can be seen in modern kossu works. If needed, the stamp sealed collector's name can also be added to the works.

缂丝纯书法作品，存世的大多于乾隆年间制作，模仿书法原作，并缂织印鉴多枚。而传统缂丝一般只能表现正楷书法，行书和草书甚为罕见。现代苏州缂丝在表现书法字体上，有了新的发展，尤其是行书和草书。1977年，苏州刺绣研究所用现代设计元素，设计制作了一系列现代设计感极强的缂丝作品，有缂丝服装、玩偶、婴儿鞋子、茶垫、香筒等。现在市场上的缂丝实用品形式还有缂丝靠垫、书本画册封面、缂丝扇套、卡包、香袋、名片夹、牙签套等。

Most of the calligraphy kossu works were completed during the Qianlong period and a large number of stamps were also produced at the same time. The regular script was widely used in traditional kossu works but it was rare to see a running script and cursive script. However, the pattern design of modern Suzhou kossu had made progress in calligraphy, especially running and cursive script. In 1977, a series of kossu works with modern elements, including clothing, dolls, baby shoes, tea mats and incense cans, were designed and produced in Suzhou Embroidery Research Institute. Some other practical kossu products also can be seen on the current market, for example, cushion, book cover, fan cover, card bag, scent bags, business card holder, toothpick cover, etc.

三、名家名作/Masterpieces

王茂仙（1896—1972年），又名王福坤，苏州吴县陆慕人。出身缂丝世家，其祖父、父亲都是缂丝艺人。14岁随父亲学艺，20多岁即自立门户。中华人民共和国成立后，参加1954年苏州市文联民间艺术刺绣小组，并培养房国平等徒弟。代表作为根据著名画家陶声甫的作品制作的《双鹅戏水》，高163cm，宽83cm。

Wang Maoxian (1896—1972) is a famous kossu craftsman in modern times, also known as Wang Fukun from Lumu, Wu County, Suzhou. And he was born in a kossu family. His grandfather and father were both kossu artist. He studied with his father at the age of 14 and became independent in his 20s. After the foundation of the People's Republic of China, he participated in the embroidery group of Suzhou Federation of Literary and Art Circles in 1954 and cultivated some students such as Fang Guoping. His representative work is *Two Geese Playing in the Water*, which is 163 cm high and 83 cm wide, based on the work of a famous painter Tao Shengfu.

沈金水（1883—1968年），苏州蠡口人，15岁从宗叔学习缂丝技艺，认真钻研缂丝技艺，织品细腻、挺括，经常被指定做织造府的订货。1954年加入苏州市文联民间艺术组刺绣生产小组，从事缂丝技艺的研究工作。缂织的题材有山水、翎毛、花卉等。收藏于故宫博物院和天津艺术博物馆的金地缂丝《牡丹》是其代表作，如图1-7-3所示。花蕊用色简练大胆，特别是花蕊的正瓣常用深色线提色，使花朵显得精神抖擞。枝干用长短戗技法不规则地织出阴阳面，粗干采用包心戗技法，即两边用色深线，中间用色浅线，外框用深色线，表现了枝干浑厚、结实的质感。

Shen Jinshui (1883—1968) is from Likou, Suzhou. He started to learn kossu techniques from his uncle at the age of 15 and he was skilled in exquisite and crisp fabrics. So he was often assigned to make clothes for the weaving house. In 1954, he participated in the embroidery production group of Suzhou Federation of literary and Art folk art group and engaged in the research of kossu technology. The pattern themes he designed included landscape, feathers, flowers, etc. *Peonies* with golden setting, collected in the Palace Museum and Tianjin Art Museum now, was his representative work, as shown in

图 1-7-3　金地缂丝《牡丹》
Kossu *Peonies* with Golden Setting

Figure 1–7–3. The color of bud was simple and novel, especially dark color was applied to express the energy of flowers. The light and shade of branches were expressed with the long–and–short draw weaving in an irregular way and the expression of the thick trunk adopts heart–wrapped draw weaving, which meant that the color on both sides was deep and the middle was light, showing the strong and solid of the branches and trunks in texture.

王金山（1939—2020 年），1956 年进入苏州刺绣工艺美术生产合作社——后更名为苏州刺绣研究所，拜著名缂丝艺人沈金水为师，开始学习缂丝工艺。经过刻苦学习，技术突飞猛进，成为中国新一代缂丝艺人中的佼佼者。独立完成的第一幅作品是以宋代名画为蓝本制作的《白头翁·竹雀》，图案准确、色泽丰富、层次分明、形态逼真，一连制作 10 幅，一销而空。为了制作优秀的缂丝作品，王金山拜师学画，为后来缂丝技艺的提高打下了坚实的基础。王金山成立"王金山大师工作室"，坚持以传统的手工技艺复制了宋代《翠羽秋荷》、朱克柔《莲塘乳鸭图》等作品，积累了丰富的经验，为后人留下了宝贵的财富。王金山被评为"国家级非物质文化遗产项目苏州缂丝制造技艺的代表性传承人"。

Wang Jinshan (1939—2020) joined the Suzhou Embroidery Arts and Crafts Production Cooperative Association in 1956 (it was renamed Suzhou Embroidery Research Institute later) and began to learn the kossu techniques by taking Shen Jinshui as his teacher. After a period of hard study, he made a great progress in making kossu fabrics and soon became the excellent craftsman of China in new generation. The first work he completed independently was *Pulsatillas and Sparrows*, which was based on the famous painting in the Song Dynasty. With exquisite pattern, rich color, clear layer and vivid image, ten pieces of his work were sold out in a short time. In order to make exquisite kossu fabrics, Wang Jinshan started to learn painting, which laid a solid foundation for the improvement of kossu technology. Later, He established Wang Jinshan studio and adhered to copy the works with traditional crafts, such as *Kingfisher and Autumn Lotus* in the Song Dynasty and *Suckling Ducks in the Lotus Pond* by Zhu Kerou, which had accumulated rich experience for kossu industry and left valuable treasure for future generations. Wang Jinshan was named as the representative inheritor of Suzhou kossu fabrics technology, one State-Level Non-Material Cultural Heritage project.

◎ 思考题/Questions for Discussion

1. 为了保护和传承缂丝文化，中华人民共和国成立后采取了哪些措施？/What measures have been taken to protect and inherit the kossu culture after the foundation of the People's Republic of China?

2. 近代缂丝创新的织造技法有哪些？/What are the innovative techniques of kossu industry in modern times?

○ 第二章
缂丝生产工艺
Production Technology of Kossu Fabrics

◎ **概述/Introduction**

通过第一章的学习，大家对缂丝的历史渊源有了一定的了解，本章将进一步介绍缂丝生产工艺。分别从缂丝原料的分类、不同原料缂丝产品的特点、经纬线特征等方面进行介绍；通过了解缂丝织机的构造及辅助工具，掌握缂丝的工艺流程；介绍缂丝的技法及特点，通过对比，深入了解缂丝工艺。

Through the study of the first chapter, I believe that everyone has a certain understanding of the historical origin of kossu. This chapter will further introduce the production process of kossu. The classification of kossu raw materials, the characteristics of kossu products of different raw materials, and the historical characteristics of warp and weft threads will be introduced; the technological process of kossu will be comprehended through understanding the structure and auxiliary tools of kossu looms; the techniques and characteristics of kossu will be introduced and the kossu craft can be deeply understood through comparison.

◎ 思维导图/Mind Map

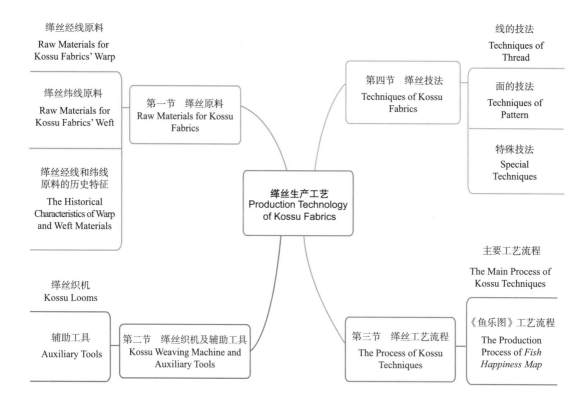

◎ 教学目标/Teaching Objectives

知识目标/Knowledge Goals

1. 掌握缂丝原料种类/The Raw Materials of Kossu Fabrics

2. 掌握缂丝生产工艺流程/The Production Process of Kossu Fabrics

3. 掌握缂丝各种技法及其特点/The Techniques and Characteristics of Kossu Fabrics

技能目标/Skill Goals

1. 能分析缂丝原料种类/Analyzing the Types of Raw Materials for Kossu Fabrics

2. 能进行缂丝生产操作/Operating the Production Process of Kossu Fabrics

3. 掌握缂丝不同技法，并能进行实操/Mastering Different Techniques of Kossu Fabrics and Carrying Out Practical Exercises

素质目标/Quality Goals

1. 具备良好的审美能力/Possessing Good Aesthetic Ability

2. 树立创新意识和创新精神/Establishing Innovative Consciousness and Spirit

思政目标/Ideological and Political Goals

1．树立实事求是的历史唯物主义观/Establishing a Historical Materialism View of Seeking Truth from Facts

2．弘扬缂丝非物质文化遗产的工匠精神/Promoting the Craftsman Spirit of the Intangible Cultural Heritage of Kossu Techniques

第一节　缂丝原料/Raw Materials for Kossu Fabrics

一、缂丝经线原料/ Raw Materials for Kossu Fabrics' Warp

缂丝所用的纱线分为经线和纬线。经线一般用没有经过练染的桑蚕丝（生丝），如图2-1-1所示，两根或三根单丝为一股，分左右两股，并用它打捻出左右方向的两股丝，通过并线合为一股，即2×4.67tex或3×4.67tex经线，如图2-1-2所示。唐、两宋以及元以后的部分经线都用强捻合线，明清以后经线捻度越来越小。

缂丝原料

The yarns used in kossu fabrics are divided into warp yarns and weft yarns. Generally, the warp is made of unscoured mulberry silk (raw silk), as shown in Figure 2-1-1. There are two or three monofilaments in one strand, which are divided into left and right strands. Two strands in the left and right directions are twisted with it and combined into one strand through parallel threads, that is, 2 × 4.67 tex or 3 × 4.67 tex warp, as shown in Figure 2-1-2. Strong twist lines were used in some warp lines after the Tang, Song and Yuan Dynasties, while the twist of warp lines became smaller and smaller after the Ming and Qing Dynasties.

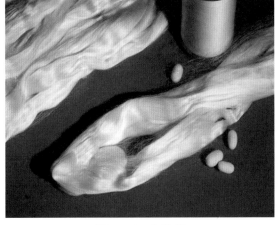

图2-1-1　生丝线
Raw Silk

图2-1-2　3×4.67tex生丝
3×4. 67tex Raw Silk

二、缂丝纬线原料/Raw Materials for Kossu Fabrics' Weft

缂丝的纬线主要有丝线、金线、孔雀羽线、麻线和毛线。

The weft threads of kossu fabrics mainly include silk thread, gold thread, peacock feather thread, hemp thread and knitting wool.

（一）丝线/Silk Thread

用于纬线的丝线分为花线、合花线、劈线。花线即是色线，如图2-1-3所示，精练后并和为丝线。一般采用经过练染加工的丝线，比经线粗，一般八根单丝为一股，分左右两股，并用它打捻出左右方向的两股丝，通过并线合为一股，即8×4.67tex纬线，根据需要增减单丝根数，合并后的纬线通常为弱捻或无捻。

The silk thread used for weft thread is divided into flower thread, mixed flower thread and split thread. Flower thread is the colored thread, as shown in Figure 2-1-3. After scouring, it is merged into Silk thread. Generally, the silk thread, processed by scouring and dyeing, is thicker than the warp thread. With eight monofilaments forming a strand, the silk thread includes two strands divided into left and right strands. In general, it is used to twist two strands of silk in the left and right directions and then combine them into one strand by doubling, that is, 8 × 4.67 tex weft thread. The number of monofilaments is increased or decreased as required and the combined weft thread is usually weakly twisted or untwisted.

合花线是将不同色彩的丝线并成一股丝线，通常是两种色线的合并，用于表现色彩的自然变化，北宋时期已经使用。南宋用于摹缂绘画作品中，例如朱克柔作品《莲塘乳鸭图》中使用较多的合花线，明清时期比较普遍，如图2-1-4所示。劈线是把单股丝线再劈成若干股，用于表现细部线条，宋代已经出现。多用于人物的胡须、眉毛以及水纹、鸟羽等，表现毛发纤细柔软的特性。

图2-1-3 色线
Colored Threads

The composite thread is a combination of silk threads with different colors into one thread, usually a combination of threads with two colors, which is used to show the natural changes of colors and had been used in the Northern Song Dynasty. In the Southern Song Dynasty, it was used to copy paintings, such as Zhu Kerou's *Suckling Duck in the Lotus Pond*, and more common in the Ming and Qing Dynasties, as shown in Figure 2-1-4. Splitting thread is a single silk thread split into several strands to show detailed lines, which appeared in Song Dynasty. It is mostly used for a character's beards, eyebrows or water waves, feathers, etc., showing the characteristics of slender and soft hair.

 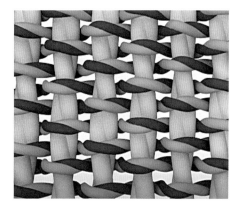

(a)《莲塘乳鸭图》的合花线　　　　　　(b)合花线示意图
The Composite Thread of *Suckling Duck in the Lotus Pond*　Schematic Diagram of the Composite Thread

图2-1-4　合花线
The Composite Thread

（二）金线/Gold Threads

缂丝用金线有片金线和捻金线两种。纺织用金线的加工工艺比较复杂：首先将金熔化，凝为片状，将金片捶打成金叶，使之变硬，经退火处理后，再垂打成金箔，然后在羊皮或纸上刷上鱼胶，粘贴上金箔，并根据不同的要求切成粗细不等的金线，即可使用，所谓片金线。根据所用材料不同可分为纸背片金和羊皮金。纸背片金缂丝出现于唐代，在宋、辽时期使用。斯坦因从敦煌盗去，现藏英国大英博物馆的缂丝遗物中发现了纸背片金缂丝。羊皮金是在薄羊皮上粘贴金片，可以整块使用，也可以切成细缕织入织物中。唐宋时期已经使用羊皮制造金箔，明清时期尤其盛行。叶梦珠《阅世篇》卷八："命妇之服……有刻丝、织纹。领袖襟带，以羊皮金镶嵌。"

There are two kinds of gold thread for kossu fabrics: sliced gold thread and twisted gold thread. The processing technology of textile gold thread is complicated: first, the gold should be melted and condensed into sheets, and the gold sheets are beaten into gold leaves until it hardens. After annealing processing, it is draped into the gold foil. Then fish glue is brushed on the sheepskin or paper, and the gold foil is pasted, and next gold threads with different thicknesses are cut for use according to different requirements, so-called sliced gold threads. According to the different materials used, it can be divided into paper-back sliced gold and sheepskin gold. Paper-backed Sliced golden kossu fabrics appeared in the Tang Dynasty and were used in the Song and Liao Dynasties.Stein stole it from Dunhuang and then the paper-back sliced gold silk was found in the kossu fabric relics in the British Museum. Sheepskin gold is a slice of gold pasted on the thin sheepskin, which can be used as a whole or cut into fine strands weaving into fabrics. Sheepskin was used to make gold foil in the Tang and Song Dynasties, especially popular in the Ming and Qing Dynasties. As recorded in the eighth Volume of Ye Mengzhu's *On Life*: "The clothes of the noble women...with kossu fabrics and woven patterns. The leader's clothes and belts are inlaid with sheepskin gold."

捻金线，也称"圆金"，在棉纱、蚕丝上涂上胶水，用片金线缠绕搓捻于外围，使之成为圆金线，如图2-1-5所示。出现于唐代，宋代使用比较多，但唐宋时片金线的使用比例更大。元代开始大量使用捻金线，明清时期则以捻金线为主，片金线较少使用。清代捻金线按照材料的种类分为赤圆金线、淡圆金线和银线。按照使用的种类又分为两色金和三色金。两色金，是指泛红光的赤金及微带青光的黄金，即赤圆金线、淡圆金线。三色金，是指赤圆金线、淡圆金线和银线。清代缂丝装饰风格华丽繁缛，往往在一件作品上使用两色或三色金。

Twisted gold thread, also known as "round gold", is coated with glue on cotton yarn and natural silk, and twisted around the periphery with a slice of gold thread to make it into round gold thread, shown in Figure 2-1-5. It appeared in the Tang Dynasty and was used more in the Song Dynasty, but with a larger proportion used in the Tang and Song Dynasties. Twisted gold thread began to be widely used in Yuan Dynasty, while during the Ming and Qing Dynasties, twisted gold thread was mainly used but less use of sliced gold thread. Twisted gold thread in the Qing Dynasty was divided into red round gold thread, light round gold thread and silver thread according to the types of materials. According to the types of use, it was divided into two-color gold and three-color gold. Two-color gold referred to red gold and yellow gold with light green, that is, red round gold thread and light round gold thread. Three-color gold referred to red round gold thread, light round gold thread and silver thread. In the Qing Dynasty, the decorative style of kossu fabrics was gorgeous and complicated, as two-color or three-color gold was often used ina work.

图2-1-5 金线
Gold Thread

例如，缂金加绣《山庄人物图》挂屏，清乾隆时期，长118cm，宽78cm。此挂屏心主要是以金线缂织山庄人物。画面中山、树、房屋等用金线缂织而成，局部用其他色线。画面在蓝色的衬托下金光异彩，富丽堂皇，如图2-1-6所示。

For example, Hanging Screen of Golden Kossu Tapestry of *Villagers* was made in the Qianlong period of the Qing dynasty, with 118 cm long and 78 cm wide. The center of this hanging screen was mainly made of villa and figure woven by a golden thread. Among which, the mountain, trees, houses, etc. were all woven with gold thread, and other color threads were also used in some parts. The picture is very splendid against the blue background, as shown in Figure 2–1–6.

图2-1-6 缂金加绣《山庄人物图》挂屏
Hanging Screen of Golden Kossu Tapestry of *Villagers*

（三）孔雀羽线/Peacock Feather Threads

孔雀羽线是将一根孔雀的羽毛与一根染成绿色的丝线捻合在一起的特殊装饰线，如图2-1-7所示。西晋时期已经出现了孔雀羽线制成的服装，唐代王公贵族用它来织造裙子、马鞍线。明清时期在上流社会中比较普遍使用。目前发现最早的掺入孔雀羽线织成的缂丝是北京定陵出土的万历皇帝的缂丝衮服，衮服上的十二团龙是用孔雀羽线织成的。清代高级补子上常使用孔雀羽线。孔雀羽线是非常珍贵的材料，价格昂贵，制作方法也很烦琐。因为孔雀羽毛短，韧性差，所以与丝线缠绕在一起使用，捻时一边捻，一边使孔雀羽毛一根接一根，连续不断。为了防止虫蛀及霉变，需将孔雀羽毛在锅中蒸煮15分钟，即可使用。

Peacock feather thread is a special decorative thread that is twisted by a peacock feather and a silk thread dyed green, as shown in Figure 2–1–7. Clothing made up of peacock feather thread appeared in the Western Jin Dynasty. It was used by nobles in the Tang Dynasty to weave skirts and saddle thread and more widely used in the upper class during the Ming and Qing Dynasties. At present, it was found that the earliest silk woven with peacock feathers was the kossu fabric robes of Emperor Wanli unearthed in Dingling, Beijing, and the twelve dragons on the robes were woven with peacock feather thread. Peacock feather thread was often used on high-level mending in the Qing Dynasty. It is a kind of very precious material, which is expensive and complicated to make. Because peacock feathers are short and have poor toughness, they are used together with silk threads. When twisting, peacock feathers are made one by one and continue continuously. In order to prevent moth-eaten and mildew, peacock feathers need to be cooked for 15 minutes before they can be used.

图 2-1-7 孔雀羽线
Peacock Feather Threads

(四) 麻线/Hemp Thread

麻纤维粗，柔软性差，与丝线混织时需要加工得非常精细。宋代缂丝中出现部分使用麻线的作品。宋代缂丝《紫芝仙寿》《翠羽秋荷》均有麻线。

As hemp fiber is coarse and has poor softness, it needs to be processed very finely when mixed with silk thread. In the Song Dynasty, there were some kossu fabrics using hemp thread. In the Song Dynasty, *Purple Ganoderma for Longevity* and *Kingfisher and Autumn Lotus* all were used with hemp thread.

《紫芝仙寿》如褐地设色织，土坡激流，石旁杂植海棠、雁来红、野菊。右岸草丛之间灵芝杂生。一株紫薇盛开，花团锦簇，双鸟栖息在枝头上，更显生意盎然。从花卉的名称及意涵来看，皆具吉祥象征，如海棠喻"满堂富贵"，雁来红有"老来娇""老少年"之名，用来送礼祝寿最合宜。此幅所用纬线材料十分别致，除背景、花朵、泉水及部分的鸟羽以丝线织成外，其余皆用麻纤维缂织而成，色彩逐层变化，过渡自然，尤其将海棠叶粗糙的质感表现得写实逼真，也能增强作品中的石块、土坡、树叶及树干厚实的量感，如图 2-1-8 所示。

Purple Ganoderma for Longevity was yarn-dyed in brown ground, with rapids on soil slopes and mixed plants of Chinese flowering crabapple, Amaranthus and wild chrysanthemum planted beside stones. Ganoderma was mixed among the grass on the right bank. Crape myrtle was in full bloom and full of flowers, and two birds inhabited the branches, which were fuller of business. From the name and meaning of flowers, they all had auspicious symbols. For example, begonia meant "full of wealth", and tricolor amaranth was also known as "Graceful old Lady" and "Mature

Young Man", which was perfect for giving gifts and celebrating a birthday. The weft thread material of this picture was very unique. Except for the background, flowers, spring water and some bird feathers were woven with silk thread, and the rest were woven with twine. The color changed layer by layer with natural transition. Especially, the rough texture of Chinese flowering crabapple leaves was vivid and realistic, and it could also enhance the weight sense of stones, soil slopes, leaves and trunks of the work, as shown in Figure 2-1-8.

（五）毛线/Knitting Wool

毛线通常指羊毛纺成的纱线，易染色，所以能织出色彩华丽的织物，具有很强的观赏效果。考古学家们在新疆发现了很多用毛线织造的缂织物。羊毛纤维短，易于松散和纠缠，经线的捻度较大，密度较小，上机必

图 2-1-8　缂丝《紫芝仙寿》
Kossu *Purple Ganoderma for Longevity*

须绷紧；而纬线要绕着紧绷的经线，只能弱捻，使纤维蓬松，保持较大程度的屈曲，排列致密，因此织物呈现纬面凸纹。

Knitting wool, made from wool, is suitable for dyeing, so it can weave colorful fabrics with a strong ornamental effect. Archaeologists have found many woven fabrics made from wool in Xinjiang. Wool fiber is short, easy to loose and entanglement. Warp twist is large, density is small, and the machine must be tightened. The weft yarn should be wound around the tight warp yarn, which can only be weakly twisted, making the fiber fluffy, keeping a large degree of buckling, and dense arrangements, so the fabric presents convex lines on the weft surface.

三、缂丝经线和纬线原料的历史特征/The Historical Characteristics of Warp and Weft Materials

从目前馆藏在博物馆的缂丝文物对比来看，宋代经线较粗，因此经线密度相对少，如宋代时经线密度在15~30根/cm，其中25根/cm以上的极少，大部分在20根左右，17~20根/cm的比例更大。明代时经线密度则普遍在25~30根/cm。元代时经线密度在13~34根/cm，大多在19~23根/cm。无论宋代还是元代，实用性作品比观赏性作品的经线密度更小，也就是说实用性缂丝的经线相对更粗。从整个经线的密度来看，宋代与元代几乎没有什么差别，

尤其是实用性作品。观赏性缂丝从比例上看，30根/cm左右的经线稍多些。

According to the comparison of kossu fabric relics collected in the museum today, the warp in the Song Dynasty was thicker, so the number of warps per centimeter was relatively small. For example, the density of warps in the Song Dynasty was 15–30 threads/cm, of which few were above 25 threads/cm, while most of which were about 20 threads/cm, and the proportion of 17–20 threads/cm was larger. In the Ming Dynasty, it was generally 25–30 threads/cm. In the Yuan Dynasty, the density of warps ranged from 13 to 34 threads/cm, and most of them were 19 to 23 threads/cm. No matter in the Song or Yuan Dynasty, the density of practical works was smaller than that of ornamental works, that was to say, the warp of practical kossu fabrics was relatively thicker. From the density of the general warp, there was almost no difference between the Song Dynasty and Yuan Dynasty, especially in practical works. In terms of proportion, there were slightly more warps of about 30 threads/cm for ornamental kossu fabrics.

宋代纬线密度以50～70根/cm为主，部分也有100根/cm以上者，如南宋缂丝艺人朱克柔的作品都在125～131根/cm。元代时纬线密度普遍在40～70根/cm。明代纬线密度集中在60～90根/cm。比较而言，实用性缂丝的纬线密度小。宋元时期的缂丝与明时期相比，纬线密度因部位不同而差别很大。这里有两个原因，一是技术不稳定，加上丝线的粗细加工的不均匀；二是技法的问题，即宋元缂丝常常在花瓣或叶子的边缘、在鸟兽的边缘用更细的、不同颜色的丝线装饰边缘。这种技法从北宋开始就有，一直到元代，在实用性和观赏性作品中都有发现，尤其在元代实用性作品中更多见。

In the Song Dynasty, the weft density was usually 50–70 threads/cm, and some of them were more than 100 threads/cm. For example, the works of Zhu Kerou, a kossu fabric artist in the Southern Song Dynasty, were all 125–131 threads/cm. In the Yuan Dynasty, it was generally 40–70 threads/cm. Wefts in the Ming Dynasty were mainly in 60–90 threads/cm. Comparatively speaking, the weft density of practical kossu fabrics was small. Compared with the Ming Dynasty, the weft density of the kossu in the Song and Yuan Dynasties varied greatly due to different parts. There were two reasons: one was that the technology was still unstable, and the thickness of the silk thread was uneven; the other was the problem of technique, that is, kossu fabrics in the Song Dynasty and Yuan Dynasty were often decorated with thinner silk threads of different colors at the edge of petals or leaves, birds and animals. This technique has existed since the Northern Song Dynasty and had been also found in practical and ornamental works, especially more common in practical works of the Yuan Dynasty.

经线加捻是唐以来的传统，在宋代，无论实用性还是观赏性缂丝的经线大多为双股强捻线，这样经线更加粗厚，以至经纬线全部织成后，在经线之间出现纵向的凸起痕迹，俗称"瓦楞地痕"。元代缂丝在实用性作品中瓦楞地痕比较普遍，但在观赏性作品中则很少见。这是由于元代实用性作品大多用双股强捻丝经线，而观赏性作品大多用弱捻或无捻线。

As for the twisting thread, twisting of warps had been a tradition since the Tang Dynasty. In the Song Dynasty, the warp of practical and ornamental kossu fabrics was mostly double–stranded strong twisted yarns, which made the warps thicker, so that after all the warps and wefts were woven, longitudinal raised marks appeared between the warps, commonly known as "corrugated ground marks". Corrugated ground marks were common in practical works of the Yuan Dynasty, but rare in ornamental works. This is because most practical works in Yuan Dynasty used double–stranded strong twisted warps, while most ornamental works used weakly twisted or untwisted threads.

◎ 思考题/Questions for Discussion

为什么缂丝的经线细纬线粗？/Why the warp threads of the kossu fabrics are thin and the weft threads thick?

第二节　缂丝织机及辅助工具/Kossu Weaving Machine and Auxiliary Tools

缂丝作品在缂丝织机上织造，同时需要各种工具才能完成，包括大拨子、小拨子、大梭子、小梭子、笔、镜子、胶水、鬃刷、纡管、撑样板。

缂丝织机及辅助工具

Kossu fabrics are woven on a silk loom, and various tools are needed to complete, including big plectrums, small plectrums, big shuttles, small shuttles, pens, mirrors, glue, brie brushes, winding tubes and supporting samples.

一、缂丝织机/Kossu Looms

（一）缂丝织机的结构/Structure of Kossu Looms

缂丝织机的主要机构有前轴、番头、竹筘、机头、踏脚棒、后轴（图2-2-1）。

The main mechanisms of the kossu weaving machine are the front axle, turning head, bamboo reeds, head, pedal stick and rear axle（Figure 2-2-1）.

1. 后轴/Rear Axle

织机的后轴为圆柱形，有卡槽，里面有一根钢丝和木条。后轴的作用是固定经线和绕经线。首先经线分股打结后绕成一个环状，把钢丝穿过经线，然后把钢丝放在卡槽中，加

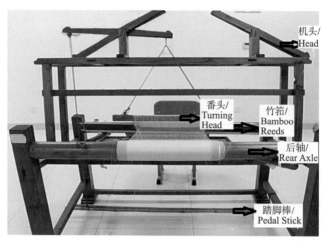

图2-2-1　缂丝织机
Kossu Weaving Machine

入木条，这样经线就固定住了。后轴的第二个作用是绕经线，经线一般10m长，在织造时，多余的经线把它绕在后轴上。

The rear axle of the loom is cylindrical and has a slot, in which there are a steel wire and a wooden strip. The function of the rear axle is to fix and wind the warp. First, the warp threads are knotted and wound into a ring, and the steel wire is passed through the warp threads. Then, the steel wire is placed in the clamping slot and some wooden strips are added so that the warp threads are fixed. The second function of the rear axle is to wind the warp thread, which is generally 10 meters long. During weaving, the extra warp threads are wound around the rear axle.

2. 番头/Turning Head

（1）番头的制作。制作一个番头时需要两块番头板，首先拿一把尺子在番头板上量取刻度，一般等于作品的宽度，画好每厘米的刻度；把一块番头板放在两个凳子中间，然后两端加一定高度的物体，一般6cm，再把刚才画刻度的番头板放上面，然后拿番头线（尼龙线）绕过番头板打结，然后开始绕圈，每厘米绕12个，一直绕到刻度最后，绕圈一定要均匀。最后拿乳胶粘一下，等干了之后把下面的板抽出来就做好一个番头了。第二个番头的制作同第一个。

Production of Turning Head. When making a turning head, you need two turning plates. First, take a ruler to measure the scale on the turn-over board, which is generally equal to the width of the work, and draw the scale per centimeter. Put a turning plate in the middle of two stools, and next add objects with a certain height at both ends, generally 6 cm, and then put the turning plate with scales on it; then take the turning line (nylon line) around the turning plate of the loom to tying a knot, and then start to circle, 12 per centimeter, until to the end of the scale, and pay attention to uniformity. Finally, take latex to stick it, and when it becomes dry, take out the following board and a turning head is done. The made process of the second turning head is in the same as the first one.

（2）番头的作用。番头和踏脚板通过绳子相连，通过脚踩竹竿，带动番头运动，番头上经线运动，形成梭口，可以穿入纬线，进行缂织。

Function of Turning Head. Turning head and pedal stick are connected by ropes, and stepping on bamboo pole with feet drives turning movement and warp movement to form shed, which can penetrate weft threads for weaving.

3. 竹筘/Bamboo Reeds

竹筘是竹子做的，上面有一片片筘片，通过筘片密度可以控制经密，一般筘片密度是24片/cm。竹筘的作用是将经线穿过筘片，通过竹筘可以控制经线密度，把纬线推向织口，还可以梳理经线（图2-2-2）。

Bamboo reeds are made of bamboo, with pieces of reeds on them. The density of warp can be controlled by the density of reeds. Generally, the density of reeds is 24 pieces per centimeter. The function of a bamboo reed is to pass the warp through the reed sheet. Through bamboo reeds, the warp density can be controlled, thus the weft can be pushed to the weaving mouth, and the warp thread can be combed（Figure 2-2-2）.

4. 前轴/Front Axle

前轴是圆柱形，分为有槽的和无槽的，其作用都是固定经线。无槽的前轴用乳胶将经线固定在前轴上，有槽的和后轴一样固定（图2-2-3）。

The front axle is cylindrical. Whether having a slot or not, front axles are all to fix warps. The front axle without slots is fixed on the front axle with latex, and the slotted front axle is fixed like the rear axle（Figure 2-2-3）.

5. 踏脚棒/Pedal Stick

踏脚棒是两根直径为3cm左右的竹竿，通过绳子将踏脚棒和番头板连接起来，脚踩踏脚棒来控制番头运动，带动经线上下运动，形成梭口。

The pedal stick is two bamboo poles with a diameter of about 3cm. The pedal stick is

图2-2-2　竹筘
Bamboo Reeds

图2-2-3　前轴
Front Axle

connected with the turning plate by a rope, and the foot tramples on the pedal bar to control the turning movement, which drives the warp thread to move up and down so as to form a shed.

（二）缂丝织机的运行原理 /Operation Principle of Kossu Weaving Machine

1. 开口运动 /Shedding Motion

缂丝织机开口运动分两种：上梭口和下梭口。脚踩踏脚棒，带动机头运动，机头连着番头板，番头板上经线上升，另一个番头板上经线不动，形成上梭口。脚踩踏脚棒，直接带动番头向下运动，形成下梭口。

There are two kinds of shedding motion of kossu loom: upper shed and lower shed. The feet step on the pedal stick to drive the head to move. The head is connected with the turning plate, and the warp on the turning plate rises, while the warp on the other turning plate does not move, forming an upper shed. Stepping on the pedal stick directly drives the turning head to move downward to form a shed.

2. 引纬运动 /Weft Insertion Motion

引纬的作用是将纬纱引入由开口机构所形成的梭口，以便经纬交织，形成织物。

The function of weft insertion is to introduce the weft yarn into the shed formed by the shedding mechanism, so that warp and weft interweave into fabrics.

有梭引纬是指由梭子将纬纱引入梭口，梭子既是引纬器，又做存纬器用。

Weft insertion with shuttle: weft yarn is introduced into the shed by shuttle, which is both a weft insertion device and a weft storage device.

（三）打纬运动 /Beating-up Motion

1. 打纬的作用 /The Function of Beating-up

用竹筘把纬纱推向织口；确定经纱密度和织物幅宽；控制梭子在梭口中的运动方向；打纬时梭子停留在梭口区域外。

Push the weft yarn to the weaving mouth with a bamboo reed; determining warp density and fabric width; controlling the moving direction of the shuttle in the shed; the shuttle stays outside the shed area during beating-up.

2. 打纬的工具 /Beating-up Tools

竹筘、拨子。

Bamboo reeds and plectrums.

3. 打纬过程 /Beating-up Process

用大、小拨子拨动纬线向梭口运动，在此过程中经纬纱间相互屈曲而产生摩擦作用，出现阻碍纬纱向织口移动的阻力，即打纬阻力。通过拨子的拨动和下一梭口打开，可以使得纬线位置相对固定。

The weft yarn is moved to the shed with large and small plectrums. In this process, the warp and weft yarns buckle each other and produce friction, which leads to the resistance that hinders the weft yarn from moving to the shed, that is, beating-up resistance. Through the plucking of the plectrum and the opening of the next shed, the position of the weft thread can be relatively fixed.

（四）卷取运动 /Take-up Motion

1. 卷取机构的作用 /The Role of the Take-up

将已织成的织物引离织口，卷绕到卷布辊上；确定和控制纬纱在织物内部的排列密度。

Leading the woven fabric away from the weaving mouth and winding it on the cloth winding roller; Determining and controlling the arrangement density of weft yarns in the fabric.

2. 卷取机构 /Tack-up Mechanism

前轴。

Front axle.

（五）送经运动 /Let-off Motion

1. 送经机构的作用 /The Role of Let-off Mechanism

能根据织物纬密的大小，从织轴上均匀送出相应长度的经纱；确定经纱所需的上机张力，并在织造过程中保持经纱张力大致稳定。

According to the weft density of the fabrics, warp yarns of the corresponding length can be evenly sent from the weaving shaft; determining the machine tension required for warp yarn, and keeping the warp yarn tension roughly stable during weaving.

2. 送经机构 /Let-off Mechanism

后轴。

Rear axle.

二、辅助工具 /Auxiliary Tools

1. 大、小梭子 /Big and Small Shuttles

大梭子作用：用于引纬，织造地部分。小梭子作用：用于织花部分引纬线，不同颜色进行换梭（图2-2-4）。

The function of the big shuttle: used for weft insertion and weaving. The Function of the small shuttle: it is used to weave weft insertion thread in flower part, and the shuttles are changed through different colors（Figure 2-2-4）.

2. 大、小拨子 /Big and Small Plectrums

大拨子作用：用于打地。小拨子作用：用于打花部分纬线。在织造时面积大的地方用大的一头，面积小的地方用小的一头（图2-2-5）。

The function of the big plectrum: it is used to hit the ground. The function of the small plectrum: it is used to beat some weft yarns. When weaving, it uses a big end for large areas and a small head for small areas（Figure 2-2-5）.

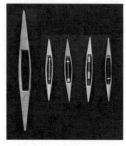
图 2-2-4　大、小梭子
Big and Small Shuttles

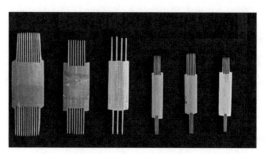
图 2-2-5　大、小拨子
Big and Small Plectrums

3. 笔、镜子和胶水 /Pen, Mirror and Glue

笔，用来画样，以前用毛笔，现在用毛笔或水笔，要求笔头细，但是水笔笔头较硬，画样没有毛笔容易，但携带方便。镜子，放在经线下面，用来随时查看织造状况。胶水，用来把经线粘在前轴上。

Writing brush was used to draw samples before, but now writing brush or fountain pen is OK. A thin pen head is required, but the fountain pen head is hard, thus, when drawing a sample, it is not as easy as a writing brush but convenient to carry. A mirror is placed under the warp thread to check the weaving condition at any time. Glue is used to stick warp threads to the front axle.

4. 鬃刷、纡管、撑样板 /Bristle Brush, Winding Tube, Supporting Template

鬃刷用于梳理经线（图 2-2-6）。纡管用于装色线的竹管（图 2-2-7）。撑样板用于放画稿。

Brie brush is used to comb warp thread（Figure 2-2-6）. Winding tube is a bamboo tube to put color thread（Figure 2-2-7）. Supporting template is used to put the drawing on it.

图 2-2-6　鬃刷
Brie Brush

图 2-2-7　纡管
Winding Tube

◎ 思考题 /Questions for Discussion

1. 缂丝织机的主要机构是什么？ /What is the main mechanism of the kossu weaving machine?

2. 缂丝织机的前轴和后轴分别的作用？/What are the functions of the front axle and the rear axle of the kossu weaving machine?

第三节　　缂丝工艺流程/The Process of Kossu Techniques

一、主要工艺流程/The Main Process of Kossu Techniques

缂丝织造技艺主要由经线上机准备→纹样稿件准备→纬线准备→打地→缂织花纹→修毛→装裱七道工序组成。

The techniques of kossu fabrics mainly consist of seven processes: Preparation of warp threads, preparation of pattern manuscripts, preparation of weft threads, Weaving the ground, weaving patterns, trimming and mounting.

（一）经线上机准备 /Preparation of Warp Threads

经线上机准备包括牵经线、穿竹箔、穿番头、嵌后轴经、拖经面、嵌前轴经和撬经面等。

The preparation of warp thread includes pulling the warp thread, threading through the bamboo reed, threading through the turning head, embedding into the warp of the rear axle, dragging the warp surface, embedding warp into the front axle and prying the warp surface, etc.

经线上机准备

牵经线：根据作品的宽度，计算经线数。一般经密为24根/cm，例如作品宽度为33cm，因此需要牵经线792根，可以适当多牵1~2根备用。根据经线数把经线牵出来，在牵的过程中分绞。注意丝线不要混乱。

Pulling the warp yarns: calculate the number of warp yarns according to the width of the work. Generally, the warp density is 24 threads per centimeter. For example, the width of the work is 33cm, so 792 warp threads are needed to be pulled, and 1–2 more warp threads can be properly pulled for later use. According to the number of warps, the warps are pulled out and twisted in the process of pulling. Be careful that the silk thread cannot be messed up.

穿竹箔：两个人配合，一个人在竹箔前，一个人在竹箔后，后面的人拿一根钩针，穿过竹箔片，然后勾上一根经线，穿过来。注意穿竹箔时不能空箔，让每根丝线都穿过箔片。

Threading through the bamboo reed: it needs two people to have a cooperation. One is in the front of the bamboo reed, the other is behind it. And the person behind takes a crochet, passing through the bamboo reed piece, and then hooking a warp thread and threading through it back.

Pay attention to not emptying the reed when threading through a bamboo reed, therefore every silk thread should thread through the reed sheet.

穿番头：穿好竹筘之后穿番头，需要两个人，一个在前，然后把番头板放在一定高度的凳子上，后面坐一个人。一个人拿着一根经线，穿过番头的环，然后另个人拉过去。注意在穿的过程中，一根穿上番头，一根穿下番头，依次穿，不能落掉。

Threading through the turning head: after threading through the bamboo reed, thread through the turning head, which requires two people's cooperation. The one in the front, puts the turning plate on a stool at a certain height and the other one sits in the other side. One person holds a warp thread and threads it through the turning head's ring. Then the other person pulls it over. Pay attention to threading one through the upper turning head, the other one through the down turning head, and threading through in turn without falling off.

嵌后轴经：把穿入筘中经线经过打结后，分别均匀嵌在后轴上。

Embedding into the warp of the rear axle: knotting the warp threads after threading through the reed, then they are evenly embedded on the rear axle respectively.

拖经面：把已经穿入筘的经线全部长度经过木梳梳匀顺后卷到后轴上。在绕轴的过程中，需要两个人配合，一个人绕后轴，一个人梳经面。绕后轴的过程中，需要加入硬纸板，放在经面下方一起绕到后轴上，从而保证一圈圈经线用硬纸板隔开。

Dragging the warp surface: comb the whole length of the warp thread that has been penetrated the reed evenly and then roll it onto the rear axle. In the process of winding, two people need to cooperate, one winding the rear axle and the other combing the warp surface. In the process of winding the rear axle, it is necessary to add a cardboard and put it under the warp surface for winding it around the rear axle, so as to ensure that every ring of warp threads is separated by cardboard.

嵌前轴经：把筘前一端经线经木梳匀嵌在前轴上。现在通常采用乳胶将经线粘在前轴上。

Embedding warp into the front axle: the warp at the front end of the reed is evenly embedded on the front axle through a wooden comb. Nowadays, latex is usually used to stick warp threads to the front axle.

撬经面：经线嵌入前轴和后轴后，通过转动后轴，将经线慢慢地卷绕在后轴上，当经线基本拉平后，再用捎桥棒将前后轴捎紧。

Prying the warp surface: after the warp is embedded into the front axle and the rear axle, the warp is slowly wound on the rear axle by rotating the rear axle. When the warp is flattened, the front and rear axles are tightened with a bridge-carrying bar.

（二）纹样稿件准备 /Preparation of Pattern Manuscripts

传统的缂丝产品种类分为生活用缂丝、观赏性缂丝和宗教用缂丝。随着历史的发展，现在缂丝主要为观赏性缂丝和生活用缂丝，以观赏性缂丝为主。目前，缂丝题材主要有人物、花鸟、书法、山水和现代抽象画等。纹样稿件准备按照作品大小要求，将作品原稿通过拍照、扫描或网上下载等方式保存到计算机，等比例放大或缩小，并打印出来。把所织纹样放在均匀平整的经面下面，用毛笔或水笔把纹样描在经面上，按样织造。

纹样稿件准备

Traditional kossu fabric products are divided into kossu fabrics for daily use, ornamental kossu fabrics and religious objects. With the development of history, now the kossu fabrics are mainly for ornamental kossu fabrics and kossu fabrics for daily use, with ornamental kossu fabrics as the main one. At present, the main themes of kossu fabrics are figures, flowers and birds, calligraphy, landscapes and modern abstract paintings and the like. Pattern manuscripts are prepared to be saved to the computer by taking pictures, scanning, or downloading online according to the size requirements of the works, and be enlarged or reduced in equal proportion, and printed out. The weaving pattern will be placed under the flat warp surface, and the pattern will be painted on the warp surface with a writing brush or fountain pen. And then weave it according to the sample.

纬线准备

（三）纬线准备 /Preparation of Weft Threads

配色摇线，依照画稿配好色线，把需要的色线分别摇在纡筒上，然后根据样子色彩把纡筒装进梭槽，即可织纬。

Match color and select threads. Match color threads according to the painting draft. Wind the required color threads on the tube respectively, and then put the tube into the shuttle groove according to the color of the draft. Hence, the weft can be used to weave.

（四）打地 /Weaving the Ground

地指的是作品最下面没有花纹的部分，打地是指将纬线与经线交织形成地部分。打地包括打地准备和具体流程。

打地、缂织花纹

The Ground refers to the bottom part of the work without patterns and weaving the ground refers to the part formed by interweaving weft and warp. Weaving the ground includes preparation and specific procedures.

首先量取打地的长度，根据地的长度再加上2~3cm，这2~3cm用于装裱。缂织工具为大拨子和大梭子。大梭子中有纡管，纡管上绕着纬线。大梭子的作用是引纬；大拨子的作用是打纬。

First, measure the length of the ground, and add 2–3 cm to the length of the base, which is used for mounting. Knitting tools include the big plectrum and big shuttle. In the big shuttle, there is a winding tube, and the weft thread is wound around it. The function of the big shuttle is to insert the weft. The function of the big plectrum is to beat up the weft.

打地的流程：脚踩前踏脚棒，前番头板向下运动带动所穿入的经线向下运动，后番头板不动，这样形成梭口，右手拿着大梭子从右向左穿过梭口，左手接过梭子，大拇指含着纬线，不能拉得紧也不能松，右手拿着大拨子打纬，要拨得均匀。脚踩后踏脚棒，后番头板向下运动带动所穿入的经线向下运动，前番头板不动，这样形成梭口，左手拿着大梭子从左向右穿过梭口，右手接过梭子，大拇指含着纬线，不能拉得紧也不能松，左手拿着大拨子打纬，要拨得均匀。

The process of weaving the ground: step on the front pedal stick, and the front turning plate moves downward to drive the warp thread it threads through moves downward, while the rear turning plate does not move, thus forming a shed. Holding a big shuttle by the right-hand to thread through the shed from right to left, and taking the shuttle in the left-hand, the thumb holds the weft thread, which cannot be pulled tightly or loosely, and the right-hand holds a big plectrum to beat up weft, which should be evenly. When stepping on the rear pedal stick, the rear flip plate moves downward to drive the warp thread move downward, while the front flip plate does not move, thus forming a shed. The left-hand holds a big shuttle to thread through the shed from left to right, and the right-hand takes over the shuttle, the thumb holding the weft line, which cannot be pulled tightly or loosely, and then the left-hand holds a big plectrum to beat up weft, which should be evenly.

（五）缂织花纹/Weaving Patterns

缂织花纹是指在缂织过程中根据花纹的颜色变化进行换纬。缂织前需要准备好各种纬线，根据作品要求配色摇线，并用小纤管分别装入小梭子。

Weaving patterns refers to changing weft according to the color change of pattern during weaving. Before weaving, you need to prepare all kinds of weft threads, wind the threads in color matching according to the requirements of the works, and put small shuttles into small winding tubes respectively.

缂织过程：由于缂丝不通梭，所以遇到不同颜色时先不织它，即"先留其处"，回过头来再织，这是缂丝与其他织物的区别所在。"杂色线"是指各种色彩的纬线，因为是小面积纯手工穿纬，所以与其他织物的"织"不同，在这里用"缀"。缂织好一种颜色需要绕一圈打结，否则在修毛时会出现大孔隙。

Weaving process: due to the kossu fabrics being not threading through the shuttle, it is not woven when meeting different colors, that is, "leave it there", and then weave it later, which is the difference between the kossu fabrics and other fabrics. "Shaded yarn" refers to weft threads of various colors, which are different from the "weaving" of other fabrics because they are handmade in small areas. Weaving a color well requires knotting after winding it a circle, otherwise, large pores will appear during trimming.

（六）修毛/Trimming

作品完成后，把正面的毛头修剪干净，使图案两面相同。另外，查看是否有漏织的地方，可以用小梭补织或用笔补色。

After finishing the work, trim the yarn ends on the front to make both sides of the pattern identical. In addition, check whether there is a weaving problem. If any, you can use a small shuttle to make up for the weaving or use a pen to make up for the color.

（七）装裱/Mounting

装裱有很多种方法。可以有类似于装裱双面绣插屏的方法，也可软裱、装框。考虑到缂丝作品的价值，应当选择合适的装裱材料，如酸枝、紫檀、黄花梨、鸡翅木等，装裱后一件完整的缂丝作品就已完美呈现。

There are many ways for mounting. There can be a method similar to mounting double-sided embroidery screen insertion, or soft mounting and frame mounting. Considering the value of kossu fabrics, appropriate mounting materials should be selected, such as mahogany, rosewood, yellow pear wood, wenge, etc., and then a complete kossu fabric work has been perfectlypresented.

二、《鱼乐图》工艺流程/The Production Process of *Fish Happiness Map*

《鱼乐图》原作为八大山人的名画。此缂丝作品制作过程包括经线上机准备、纹样稿件准备、画样、纬线准备、打地、缂织花纹、修毛和装裱。根据作品颜色，这幅作品为水墨画，因此需要不同深浅的灰色以及黑色共9种颜色的纬线。缂织时用到的技法有勾、结、平缂、搭梭、戗色和子母经等。其中子母经主要用于缂织图章用，在书画作品中十分常见（图2-3-1~图2-3-9）。

Fish Happiness Map was originally a famous painting of the Bada Shanren. Its production process includes preparation of warp threads, preparation of pattern manuscripts, draft drawing, preparation of weft threads, weaving the ground, weaving patterins, trimming and mounting. According to the color of the work, this work is an ink painting, so the weft threads need 9 shades of gray and black in different shades. The techniques used in weaving include hook weaving, knotting weaving, flat weaving, shuttling weaving, Qiang Se and Zi Mu Warps weaving. Among them, Zi Mu Warps Weaving is mainly used for weaving seals, which is very common in paintings and calligraphy works（Figures 2-3-1 to 2-3-9）.

图2-3-1　经线上机准备
Preparation of Warp Threads

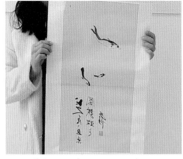

图2-3-2　纹样稿件准备
Preparation of Pattern Manuscripts

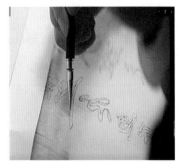

图2-3-3　画样
Sample

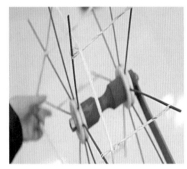

图2-3-4　纬线准备
Preparation of Weft Threads

图2-3-5　纬线绕线
Weft Winding

图2-3-6　打地
Weaving the Ground

图2-3-7　缂织花纹
Weaving Patterns

图2-3-8　修毛
Trimming

图2-3-9　装裱
Mounting

◎ **思考题/Questions for Discussion**

1. 缂丝织造流程中，纬纱准备包括哪些？/What are the weft preparations included in the weaving process of kossu fabrics?

2. 阐述缂丝的织前准备。/Explain the preparation before weaving kossu fabrics.

第四节　缂丝技法/Techniques of Kossu Fabrics

缂丝的技法较多，主要包括线的技法、面的技法和特殊技法。

There are many techniques of kossu fabrics, including techniques of thread, pattern and specialty.

缂丝技法

一、线的技法/Techniques of Thread

缂丝织物中，常见的图案线条如花瓣的边缘、叶子的叶脉等，用勾、绕等技法进行勾勒。

In the kossu fabric, the stripes of common patterns, such as the edges of petals and the veins of leaves, are outlined by hooking and winding.

1. 勾/Hooking

纹样外缘一般均用较本色深的线清晰地勾出外轮廓，如同工笔勾勒般的效果。勾缂，也称构缂，技法出现于唐代，宋、元、明、清时期一直使用。常用于细线条，多为轮廓线和结构线，如花、叶等边缘用另一色纬缂织，使花纹界限清楚，如图2-4-1所示。勾分单勾（以单股丝缂）和双勾（以双股丝缂）或多层勾。例如，《木槿花图》中花瓣与绿叶则采用合花线织出色阶的变化，花叶的勾勒线也是断断续续，表现出织者的独特创意。

Generally, the outer edge of the pattern is clearly outlined the outer contour with a deeper color thread, which has the same effect as meticulous brushwork. Hooking kossu, also known as structure kossu, appeared in the Tang Dynasty and was also used in the Song, Yuan, Ming and Qing Dynasties. It is commonly used for thin lines, mostly contour lines and structural lines, such as the edges of flowers, leaves and others woven with another color weft, so that the pattern boundaries are clear, as shown in Figure 2–4–1. The hooking is divided into a single hooking (with the single thread) and a double hooking (with double thread) or multi–layer hooking. For example, the petals and green leaves of *the Hibiscus Flower* are woven with a combination of mixed colors lines, and the outline lines of flowers and leaves are intermittent, showing the unique creativity of the weavers.

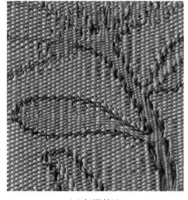

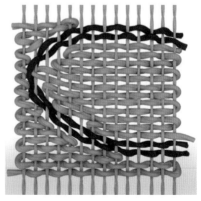

| (a) 勾缂技法 | (b) 勾缂技法示意图 |
| Hooking Skills | Schematic Diagram of Hooking Weaving |

图2-4-1　勾缂
Hooking

2. 绕/Winding

在一根或几根经线上，单梭绕出直斜、弯曲的各种线条，织成后看有镶嵌般的效果。宋代已经使用，明清时期普遍使用，且手法纯熟。

A single shuttle winds out various straight, oblique and curved lines on one or several warp threads, which has a mosaic effect after weaving. It had been used in the Song Dynasty and was widely used in the Ming and Qing Dynasties, and its technique was skillful.

3. 盘梭/Winding Shuttling

通常在一根经线或平面上，用两只梭子交织的方法循环往复，一直是两梭相交织的叫盘梭。按照同样方法，在两根经线上进行交织线条比一根粗，如此织造法的线条立体效果较好。

Two shuttles interweaving are utilized to reciprocate on a warp thread or plane in general, which is called winding shuttling. According to the same method, interweaving lines on two warp threads are thicker than one, so the stereo effect of the weaving method is better.

二、面的技法/Techniques of Pattern

缂丝作品中大部分是通过各种技法形成图案的面，其中包括平缂、掼缂、结缂、长短戗、木梳戗、凤尾戗、包心戗等十多种技法，不同的技法用于表现不同的题材。

Most of the kossu works form patterns through various techniques, including more than ten techniques, such as flat weaving, Guan weaving, knotting weaving, long–and–short draw weaving, wooden–comb draw weaving, phoenix tail draw weaving, and heart–wrapped draw weaving. Different techniques are used to express different themes.

（一）平缂 /Flat Weaving

平缂是缂丝基本技法，依照图案色彩的变化要求顺经纬之理进行平纹交织。常用于背景底色和小型图案，如图2-4-2所示。

Flat weaving is the basic technique of kossu fabric, that means, the plain weave is interwoven with the principle of longitude and latitude according to the change of pattern color. This method is commonly used for background color and small patterns, as shown in Figure 2–4–2.

（二）掼缂 /Guan Weaving

在一定坡度的纹样中（除单色外），二色以上按色之深浅有规律、有层次地排列，如同叠上去似的和色方法，如图2-4-3所示。多用于鸟的翅膀、花瓣等彩色变化位置，表现自然的色彩组合和晕色转变。唐代已经开始使用，主要用于山石、云层的装饰。

In a pattern with a certain slope (except monochrome), more than two colors of the pattern are arranged regularly and hierarchically according to the depth of colors, as if they are stacked up with mixing colors, as shown in Figure 2–4–3. It is mostly used for positions with color change

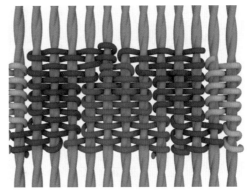

（a）平缂技法
Flat Weaving Skills

（b）平缂技法示意图
Schematic Diagram of Flat Weaving

图2-4-2　平缂
Flat Weaving

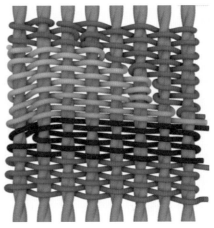

（a）搅缂技法
Guan Weaving Skills

（b）搅缂技法示意图
Schematic Diagram of Guan Weaving

图2-4-3　搅缂
Guan Weaving

such as wings of birds and petals, showing natural color scheme and iridescence change. It has been used since the Tang Dynasty, mainly for the decoration of rocks and clouds.

（三）结缂/Knotting Weaving

单色或二色以上的纹样竖的地方采取有一定规律的面积穿经的和色方法。常用于模拟自然花鸟、云气水波等自然过渡色彩，丰富层次感，如图2-4-4所示。宋代出现，使用比较广泛，多用于山石、花瓣等。

Monochrome or more than two-color patterns in vertical stripes adopt the method of mixing colors with regular areas. It is often used to simulate natural transitional colors such as natural flowers and birds, clouds and water waves to enrich the sense of depth, as shown in Figure 2-4-4. It appeared and was widely used in the Song Dynasty, mostly for rocks and petals.

 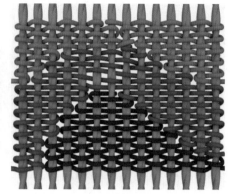

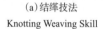

(a) 结缂技法
Knotting Weaving Skill

(b) 结缂技法示意图
Schematic Diagram of Knotting Weaving

图 2-4-4　结缂
Knotting Weaving

（四）戗色 /Draw

戗是指两种以上的色纬相邻相靠，即用两种或两种以上的色彩配合缂织的技术，称为
"戗色"。在实践过程中，艺人们根据不同图案，灵活使用相应的戗法，可以分为长短戗、
木梳戗、凤尾戗、包心戗等。

Draw refers to the adjacent weft of two or more colors, that is, the technique of weaving
with two or more colors, which is called "Draw Color". In practice, handcraft makers use the
corresponding methods flexibly according to different patterns, which can be divided into long-
and-short draw weaving, wooden-comb draw weaving, phoenix tail draw weaving, heart-
wrapped draw weaving and so on.

1. 长短戗 /Long-and-short Draw Weaving

以长短不同的各色丝线，根据物体生长的特点无规则地缂织出自然的效果。如为增
加鸟的羽毛、人物的相貌、花叶等部位的质感，在由深至浅的晕色中利用织梭伸展的长
短变化，使深色纬与浅色纬无规则地相互穿插，在视觉上产生色彩空间混合的效果，从而
取得自然晕色的印象。这种方法在宋代绘画题材中广泛使用，在南宋缂丝艺人朱克柔的作
品上熟练使用，所以也称"朱缂"法。其戗法的特点是无规格，根据图案的需要灵活变化
（图 2-4-5）。

The silk threads of different lengths are randomly woven to produce a natural effect according
to the characteristics of the growth of the object. For example, in order to increase the texture of
the feathers of birds, the appearance of the characters, the flowers and leaves, etc., the length of
the weaving shuttle is used in the color from dark to light, so that the dark color weft and the light
color weft are randomly interspersed with each other. Visually it produces the effect of color space
mixing, so as to obtain the impression of natural halo. This method was widely used in painting themes
in the Song Dynasty, and skillfully used in the works of Kerou Zhu, a silk artist in the Southern Song

Dynasty, so it was also called the "Zhu kossu" method. The characteristic of the method is that it has no specifications and can be flexibly changed according to the need of patterns（Figure 2–4–5）.

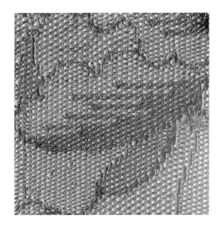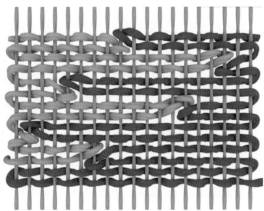

（a）长短戗技法
Long–and–short Draw Weaving Skills

（b）长短戗技法示意图
Schematic Diagram of Long–and–short Draw Weaving

图2–4–5　长短戗
Long–and–short Draw Weaving

2. 木梳戗 /Wooden–comb Draw Weaving

以深浅不同的各色长短丝线从左向右或从右向左排列成整齐的形如木梳的影光条，故名木梳戗。具有色彩渐渐过渡、色条规整的装饰效果。宋元时期普遍使用，如宋缂丝《花鸟图轴》中叶子用深浅不同的绿色长短纬线横向织出木梳的效果，以表现叶子的层次（图2–4–6）。

The long–and–short silk threads of different shades are arranged from left to right or from right to left into a neatly shaped shadow and light strip like a wooden comb, hence the name wooden–comb draw weaving. It has the decorative effect of gradual transition of colors and regular color bars. It was commonly used during the Song and Yuan dynasties, such as in the Song kossu *Flower and Bird Drawing Axis*, the leaves were woven with wooden combs with different shades of green long and short wefts horizontally to express the layers of the leaves（Figure 2–4–6）.

3. 凤尾戗 /Phoenix Tail Draw Weaving

与木梳戗的原理相同，只是凤尾戗所织出的形状不同，如图2–4–7所示。凤尾戗形状如凤凰的尾巴，故名。用来表现鸟类的羽毛或山石的阴影，宋代已经出现。

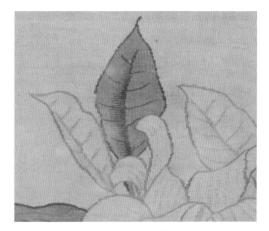

图2–4–6　木梳戗技法
Wooden–comb Draw Weaving Skill

The principle of phoenix Tail draw weaving is the same as wooden–comb draw weaving, but the woven shape is different, as shown in Figure 2–4–7. The phoenix Tail draw weaving is shaped like the tail of a phoenix, hence it's named phoenix Tail draw weaving. It appeared in the Song Dynasty used to express the feathers of birds or the shadows of rocks,

(a) 凤尾戗技法
Phoenix Tail Draw Weaving Skills

(b) 凤尾戗技法示意图
Skills Schematic Diagram of Phoenix Tait Draw Weaving

图 2-4-7　凤尾戗
Phoenix Tait Draw Weaving

4. 包心戗 /Heart–wrapped Draw Weaving

以长短戗的原理从四周同时向中心戗色，使颜色产生深浅不同的层次变化，使图案具有立体感。多用于较大面积的戗色，如鸟背、树干等，南宋时期缂丝艺术家沈子蕃常用此技法。如在沈子蕃作品《梅花寒鹊图立轴》中，鹊背部用深浅不同的黑灰色系，以参差不均地向中心包围，使羽毛生动饱满。

Based on the principle of long–and–short draw weaving, the color is changed from the four sides to the center at the same time, so that the color changes in different shades, and the pattern has a three–dimensional effect. It is mostly used for large areas of color, such as bird backs, tree trunks. This technique was commonly used by the kossu artist Shen Zifan during the Southern Song Dynasty. For example, in Shen Zifan's *Plum Blossoms amd Cold Magpies*, the back of the magpie was surrounded by different shades of black and gray to the center, making the feathers vivid and full.

5. 参合戗 /Crossing Draw Weaving

参合戗也是一种表现色彩由深到浅过渡的和色方法，但深浅两色的交替不一定绝对平均，而可以比较灵活地掌握其深浅变化的层次关系（图 2-4-8）。

Crossing Draw weaving is also a method to express the transition of color from deep to light, but the alternation between deep and light colors is not necessarily absolutely average, and the hierarchical relation between deep and unchanged can be grasped flexibly（Figure 2–4–8）.

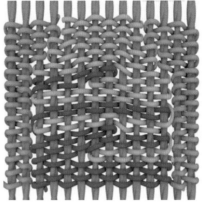

| （a）参合戗技法 | （b）参合戗技法示意图 |
| Crossing Draw Weaving Skills | Schematic Diagram of Crossing Draw Weaving |

图 2-4-8 参合戗
Crossing Draw Weaving

三、特殊技法 /Special Techniques

1. 搭梭 /Shuttling Weaving

（1）搭梭的概念。当缂织两种不同颜色的花纹边缘时，因两色不能互相连接而形成裂缝，故在每隔一定距离处，使两边的色纬相互搭绕对方色区的一根经线，以避免竖向裂缝过长而形成裂口，此种缂织技法称搭缂，如图2-4-9所示。

The concept of shuttle weaving. When weaving the edges of two patterns with different colors, cracks are formed because the two colors cannot be connected. Therefore, at regular intervals, the color weft on both sides is wound around a warp thread in each other's color area to avoid the vertical cracks being too long and forming cracks. This weaving technique is called "shuttling weaving", as shown in Figure 2-4-9.

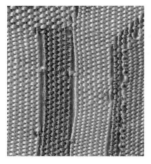
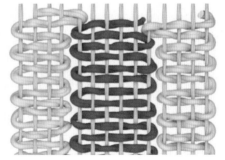

| （a）搭梭技法 | （b）搭梭技法示意图 |
| Shuttling Weaving Skills | Schematic Diagram of Shuttling Weaving |

图 2-4-9 搭梭
Shuttling Weaving

（2）搭梭技法。当两种色线相邻或遇到垂直线时，因双方不相交而有裂痕，所以在每隔一定距离处，让两边的色纬相互搭绕一次，绕过对方色区内的一根经丝。这样既能不留织纹痕迹，又能避免竖向裂缝过长，形成破口。搭梭的技法根据不同情况灵活多变，搭绕的频率不规则。在书法作品中搭梭技术普遍使用。

The shuttling weaving skills. When two kinds of color lines are adjacent or intersect with vertical lines, there are cracks because they do not meet each other. Therefore, at regular intervals, the color weft on both sides laps around each other once, bypassing a warp in each other's color area. In this way, it can not only leave no trace of weaving but also avoid the vertical cracks being too long and forming crevices. The technique of shuttling weaving is flexible and changeable according to different situations, and the frequency of it is irregular. The shuttling weaving technique is widely used in calligraphy works.

2. 子母经/Zi Mu Warps Weaving

在缂丝织造上，直线条要达到无竖缝，则在织造时运用两只梭子，即甲、乙二梭，当甲梭在墨样上穿一梭，而乙梭通穿纬线时跳过墨样一根经，让甲梭挑穿，如此原地往复，则形成无竖缝单子母织造法，如图2-4-10所示，而双子母与单子母的不同在于单子母跳过一根经线，而双子母跳过两根经线，显得比单子母粗一倍而已。

In the tape weaving, if the straight line has no vertical seams, two shuttles are used when weaving, namely the shuttle A and the shuttle B. When the shuttle A threads through the ink sample, the shuttle B runs through the weft and skips one warp thread of the ink painting sample and lets the shuttle A pick it through. In this way, reciprocating in place will form a single Zi Mu weaving method without vertical seams, as shown in Figure 2-4-10. The difference between double Zi Mu weaving and single Zi Mu weaving is that the method of single Zi Mu weaving skips one warp, while the double Zi Mu weaving skips over two warp threads, which appears to be twice as thick as the method of single Zi Mu weaving.

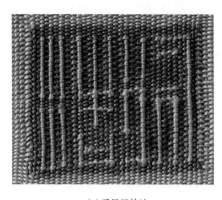
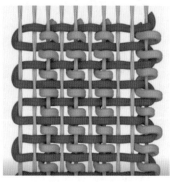

(a) 子母经技法　　　　　　　　(b) 子母经技法示意图
Zi Mu Warp Weaving Skills　　　Schematic Diagram of Zi Mu Warps Weaving

图2-4-10　子母经
Zi Mu Warps Weaving

3. 芦菲片 /Rutabaga Piece

用梭子穿经线二上二下，回梭时需排前一根经，仍穿二上二下，往回穿一段长度，再织时，每梭都需排前一根经丝穿梭又织一段长度，这样就织造出"人"字形花纹，反复织造就像一片芦菲，如图 2-4-11 所示。芦菲片多用于隔花图案，如"窗格花""门花"等，也适用于织造象芦席片的图形，但织造时需要换番头。在纺织领域，也可以把这样的结构称为纬山形斜纹。

Use the shuttle to thread the warp two up and two down, back to the shuttle need to line up the previous one, still wear two up and two down, back to wear a length, then weave each shuttle need to line up the previous one warp shuttle and weave a length, so that the "人" pattern is woven, repeatedly woven like a piece of rutabaga, as shown in figure 2-4-11. It is mostly used for the pattern of the window lattice, door flowers and so on. It is also suitable for weaving patterns that resemble pieces of reed mat, but the weaving requires a change of turning head. In the textile field, such a structure can also be referred to as a weft mountain twill.

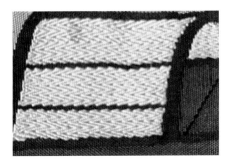 

| (a) 单色芦菲片 | (b) 双色芦菲片 |
| Single Colour Rutabaga Piece | Double Colour Rutabaga Piece |

图 2-4-11 芦菲片

Rutabaga Piece

4. 押帘梭 /Curtain Shuttle

运用甲、乙两梭线色，用甲线色在墨样上一头向下，乙线色通织二梭，甲线色在墨样上下调头，乙线色又通织二梭，甲线色再上下调头，它可以根据墨线弯曲走向调转，织出很细的直线条，如图 2-4-12 所示，可用于缂织蝴蝶、蜻蜓须、脚等图案。

The use of A, B two shuttle thread colour, with A thread colour on the ink sample head down, B thread colour through the weave two shuttles, A thread colour in the ink sample up and down, B thread colour and weave two shuttles, A thread colour and then up and down, as shown in figure 2-4-12,it can be according to the ink line bending direction to turn, weaving a very thin straight line, such as butterflies, dragon whiskers, feet, etc.

5. 押样梭 /The Pegging Shuttle

运用甲乙两只色线梭子，甲梭穿二上二下往返回梭，乙梭通穿二梭，如此循环往复，就呈现花蕊细点隆凸的效果。多用于隔花图案,如"窗格花""门花"等,如图2-4-13。

The use of A and B two colour thread shuttle, A shuttle wear two up and two down to return to the shuttle, B shuttle through the two shuttle, so the cycle repeated, to present the effect of the pistil fine point convex. This is mostly used for patterns such as "windowpane flowers" and "door flowers", as shown in figure 2-4-13.

图2-4-12 押帘梭　　　　　　　　图2-4-13 押样梭
Curtain Shuttle　　　　　　　The Pegging Shuttle

6. 缂鳞法 /The Ke Scale Method

缂鳞法是唐宋时期创造的一种技法,在明代得以发展。为明清两代的典型缂法,它是以斜戗和环戗为主,因形象似鱼、龙的鳞甲片,故称为"缂鳞法",如图2-4-14所示。该方法也可以用于缂织鸟、凤凰等的羽毛,能够使物像更加生动逼真。如明代定陵出土的缂丝龙袍上的云龙鳞甲及明代缂丝"凤穿牡丹"的凤凰羽毛片,均是采用这一缂法。

The ke scale method is a technique created during the Tang and Song dynasties and developed for the Ming dynasty. For the Ming and Qing dynasties, the typical kossu method, it is based on oblique bumping and ring bumping, because the image resembles fish and dragon scale armour pieces, so called "ke scale method", as shown in figure 2-4-14.It can also be used to kossu birds, phoenixes and other feathers, which can make the object more vivid and realistic. For example, the kossu dragon scale armour on the dragon robe unearthed in the Dingling tomb of the Ming Dynasty and the phoenix feather pieces of the Ming Dynasty kossu " Phoenix through the Peony " are all made using this kossu method.

清乾隆年间的蓝色缂丝云龙纹单朝袍（图2-4-15）,身长144cm,两袖通长194cm,袖口17cm,下摆160cm,开裾长18cm。朝袍圆领,大襟右衽,马蹄袖,披领为石青色,明黄色绦带背云,上衣下裳相连。袍以蓝色缂丝为地,用两色圆金线缂织金龙纹,龙纹处采用缂鳞法,五彩丝线织祥云、八宝平水及海水江崖等纹饰。

A blue kossu cloud and dragon motif single court robe（figure 2-4-15）, in Qianlong period of Qing dynasty, 144 cm long, both sleeves 194 cm long, cuffs 17 cm, hem 160 cm, open train 18 cm long. The robe has a rounded collar, large lapels with a right overlap, horseshoe sleeves, and a draped collar of stone blue with a bright yellow tapestries backed by clouds, the upper garment attached to the lower garment. The robe is decorated with blue kossu silk for the ground, with dichroic round gold thread kossuing gold dragon pattern, dragon pattern at the use of kossu scale method, multicoloured silk thread weaving auspicious clouds, eight treasure flat water and sea water river cliff and other decorations.

图2-4-14 缂鳞法
The Ke Scale Method

图2-4-15 蓝色缂丝云龙纹单朝袍
A Blue Kosuu Cloud and Dragon Motif
Single Court Robe

7. 削梭/Cutting Shuttle

在较宽的经面上来回穿梭不方便，那么可在每穿一梭时，在返梭时尾部留5～7cm顺梭，如同斜坡形，待接梭时就在留下的5～7cm处相衔接。这样的织造法，既解决了阔幅经面不易织造且费劲的问题，又可使整个经面在织造上不露破绽。如果是金地织底，来返缩梭还可短些，可达同样的效果。

It is not convenient to shuttle back and forth on the wider warp surface, then you can leave 5–7 cm of shuttle at the end of the return shuttle when each shuttle is worn, as in the shape of a slope, to be connected at the 5–7 cm left by it when the shuttle is picked up. This method of weaving not only solves the problem that the broad warp is not easy to weave and is laborious, but also allows the entire warp to be woven without revealing any flaws. If it is a gold ground weaving bottom, to return to the shrinkage shuttle can also be shorter, can achieve the same effect.

8. 缂绘/Kossu Painting

缂绘是基于缂丝工艺产生的一种缂织技艺，是指运用"通经断纬或者回纬"的缂丝技艺织出图案轮廓，在缂织过程中若不想完全缂织图案，则只需缂织出绘画稿本中图案的外部轮廓，图案内部不同色块处理，而是"以绘代缂"，根据绘画稿本的底色着笔描绘。

It is a kosuu technique based on the kossuing process, which means that the outline of the pattern is woven using the kossuing technique of "pass through the warp and break the weft or back to the weft", during the kossuing process if you do not want to completely Ke the pattern, you only need to kossu the external outline of the pattern in the painting draft, the processing of the different colour blocks inside the pattern is "painted instead of Ke", according to the base colour of the painting draft to depict the brush.

大红色缂丝彩绘八团梅兰竹菊袷袍（图2-4-16），这件袷袍在大红色平纹地上运用平缂、搭缂等技法缂织海水江崖纹，而八团梅兰竹菊纹样则用笔进行描绘。该袷袍设色丰富，晕色和谐，缂工细腻精湛，彩绘逼真生动。

A red kossu painted lined robe with eight clusters of plum, orchids, bamboo and chrysanthemums,（Figure 2-4-16）, This lined robe is kossued on a red plain ground using techniques such as flat kossuing and lap kossuing, while the eight groups of plum, orchid, bamboo and chrysanthemum motifs are depicted with a brush. The robe is rich in colour, with harmonious colour hues, delicate and exquisite kossu work, and vividly painted.

图2-4-16　大红色缂丝彩绘八团梅兰竹菊袷袍
A Red Kossu Painted Lined Robe with Eight Clusters of Plum, Orchids, Bamboo and Chrysanthemums

9. 缂绣/Kossu and Embroidery

缂绣即把缂丝和刺绣两者结合起来，运用到同一块织物上，这在一定程度上加强了织物的装饰效果，丰富和提高了缂丝艺术的表现力。如清乾隆年间的缂丝加绣《九阳消寒图》（图2-4-17），苏州绣制，纵213cm，横119cm。此图以缂丝加刺绣制成。其背景为缂丝，而主要人物、动物及树木等则是在缂丝上加绣。图中运用了平戗、勾边线、搭梭等缂丝技法和戗针、钉线、施针、斜缠针、打籽针、擞和针等刺绣针法，色彩丰富艳丽，特别是蓝与红、白等色的对比，使得主题突出，物象分明。

The combination of Kosuu and embroidery means that the two are combined and applied to the same piece of fabric, which to some extent enhances the decorative effect of the fabric and enriches and improves the expressive power of the kossu art. For example, the Kossu with embroidery of *Nine Suns in the Cold* (Figure 2-4-17), in Qianlong period of Qing dynasty, Suzhou embroidery, 213 cm in length and 119 cm across. The figure is made of Kossu silk with embroidery. The background is Kossu, while the main figures, animals and trees are embroidered on top of the Kossu .The kossu technique, including flat bumping, border hooking and shuttle, and the embroidery stitches, such as bumping, stapling, sizing, diagonal winding, seeding and conching, are used to create a rich and colourful scene.

图2-4-17 缂丝加绣《九阳消寒图》
The Kossu with *Embroidery Nine Suns of in the Cold*

◎ **思考题**

1. 缂丝戗法有哪些？/What are the methods of "Draw" to make kossu fabrics?
2. 戗色主要是达到什么目的？/What is the main goal of coloring of Draw?

◯ 第三章

缂丝产品类型
Product Types of Kossu Fabrics

◎ 概述/Introduction

　　本章介绍缂丝产品的分类，从服装到服饰品，详细讲述了缂丝在服饰方面的应用；通过介绍缂丝在团扇上的应用、发展及特点，使学习者掌握缂丝团扇发展的脉络和形制；同时，通过赏析缂丝名家名作，深入了解缂丝产品类型，进一步培养学习者的创新精神和工匠精神。

This chapter introduces the classification of kossu products from clothing to apparel, and describes the application of kossu in clothing in detail; it aims to let learners master the skeleton of the development of kossu fan by describing the application, development and characteristics of kossu in fan. Meanwhile, appreciating the famous kossu products of other types, and learning more about the patterns of kossu can cultivate the innovative spirit and craftsman spirit of the learners.

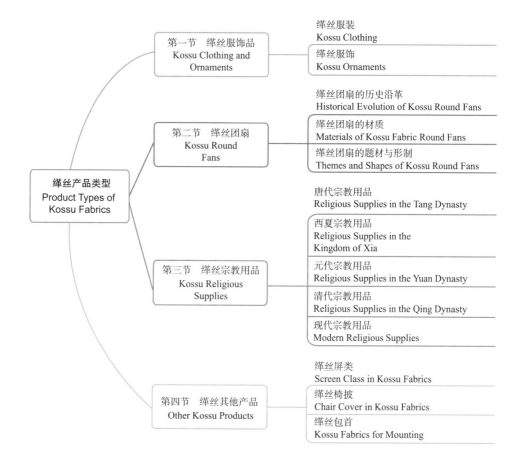

◎ 教学目标/Teaching Objectives

知识目标/Knowledge Goals

1. 了解缂丝产品的主要分类 /The Main Categories of Kossu Products

2. 掌握缂丝服饰品的种类及特点 /Types and Characteristics of Kossu Clothing Ornaments

3. 掌握缂丝团扇的发展脉络和形制 /The Development and Shape of Kossu Fabric Round Fan

4. 赏析缂丝名家名作 /Famous Works of Kossu Fabrics

技能目标/Skill Goals

1. 熟练掌握缂丝产品的主要分类 /Acknowledging the Main Categories of Kossu Products

2. 能分析不同种类缂丝的发展及特点 /Analyzing the Development and Characteristics of Different Kinds of Kossu Fabrics

素质目标 /Quality Goals

1. 具备良好的分析问题和总结问题的能力 /Possessing Good Analytical Ability and Summarizing Ability

2. 提高实践能力、培养工匠精神 /Improving Practical Ability and Cultivate Craftsmanship

思政目标 /Ideological and Political Goals

1. 培养爱国主义情操 /Cultivating Patriotism

2. 提升思想道德水平和文化素养 /Improving Ideological and Moral Level and Cultural Accomplishment

第一节   缂丝服饰品 /Kossu Clothing and Ornaments

一、缂丝服装 /Kossu Clothing

缂丝服饰品

唐代时，已有缂丝女佣束腰带，并于 1973 年出土，如图 1-1-3 所示。到了辽代，缂丝是皇室和贵族主要使用的袍服面料之一，据《辽史》记载"小祀皇帝硬帽，红缂丝龟文袍，皇后戴红帕，服络缝红袍，县玉佩，双同心帕，络缝乌靴"。

In the Tang Dynasty, there was maid girdle of kossu fabric unearthed in 1973, as shown in Figure 1-1-3. In the Liao Dynasty, kossu fabrics were one of the robe fabrics mainly used by the royal families and nobles. According to *History of Liao*, "the hard hat, the red kossu fabric turtle robe worn by the emperor, the red handkerchief, the red sewn robe, the jade pendant, the double concentric handkerchief, and the black sewn boots worn by the queen".

明代时，缂丝实用品主要为衣缎、补子和罩甲。明代缂丝中除了丝线、金线外，还使用了孔雀羽毛。嘉靖年间，缂丝袍服的制作已经有了相当的规模。

In the Ming Dynasty, the practical products made by kossu fabrics were mainly satin, buzi and covers. In addition to silk thread and gold thread, peacock feathers were also used in the Ming Dynasty. During the Jiajing period, the production of kossu dragon robes for Chinese emperors had already reached a considerable scale.

清代在乾隆时期，国力鼎盛，缂丝袍服大增，至今故宫博物院中缂丝除了书画作品外大多为缂丝衣料、朝服、吉服袍、补服等。图 3-1-1 所示为缂丝锦鸡方补，长 31cm，宽 29.5cm。补子以金线织地，上部为五彩祥云及红日，下部为海水江崖纹。中间饰一只展翅的锦鸡独立于海中礁石上，回首遥望红日。锦鸡头顶的羽毛、腿、翅膀和翎用捻毛线缂织，富有绒毛的质感。其余纹样用彩色丝线缂织而成。

During the Qianlong period of the Qing Dynasty, the national strength was at its peak, and the number of kossu robes increased greatly. Up to now, except for paintings and calligraphy, most of the kossu robes in the Palace Museum are kossu clothes, royal robes, auspicious robes and supplementary clothing robes. As shown in Figure 3–1–1, the golden pheasant with silk was 31 cm long and 29.5 cm wide. The patch was woven with gold thread, with colorful clouds and red sun on the upper part and a sea water cliff pattern on the lower part. In the middle, a golden pheasant with wings spreading was independent standing on the rocks of the sea, looking back at the red sun. The feathers, legs, wings and feathers on the top head of the golden pheasant were woven with twisted wool, which was full of fluffy texture. Other patterns were woven with colored silk threads.

图 3-1-2 所示为蓝色缂丝云龙纹单朝袍，身长 144cm，两袖通长 194cm，袖口 17cm，下摆 160cm，开裾长 18cm。朝袍圆领，大襟右衽，马蹄袖，披领为石青色，明黄色绦带背云，上衣下裳相连。袍以蓝色缂丝为地，用二色圆金线缂织金龙纹如制，五彩丝线织祥云、八宝平水及海水江崖等纹饰，衣身列十二章。此朝袍缂工细密，正反两面缂织均平齐匀整，不露线头。颜色以红、深绿、浅绿、深蓝、浅蓝、淡紫等色为主，色彩丰富，层次分明。根据清代典制，此蓝色朝袍应是乾隆皇帝在南郊祈谷、雩祭场合时所穿用。

Blue kossu imperial robe with dragons among polychrome clouds, as shown in Figure 3–1–2, had a length of 144 cm, a sleeve length of 194 cm, a cuff length of 17 cm, a hem of 160 cm and a train length of 18 cm. The robe had a round neck, a big lapel and a right lapel, horseshoe sleeves, and the collar was stone cyan, with bright yellow sash and clouds on the back, and the coat was

图3-1-1　缂丝锦鸡方补
Kossu Fabric Golden Pheasant Recipe
Supplement

图3-1-2　蓝色缂丝云龙纹单朝袍
Blue Kossu Imperial Robe with Dragons Among
Polychrome Clouds

connected with the underdress. The robe was made of blue kossu fabrics, and the golden dragon pattern was woven with two–color round gold thread. The colorful silk thread was woven with auspicious clouds, eight treasures, flat water and sea water and river cliffs, etc. The body of the robe was listed in twelve kinds of ornamentation. This robe was carefully woven, and the front and back sides were evenly woven, without revealing the thread ends. The colors were mainly red, dark green, light green, deep blue, light blue and lavender, with diverse colors and distinct levels. According to the Qing Dynasty canon system, this blue robe should be worn by Emperor Qianlong when praying for rich and ripe grains as well as rain.

二、缂丝服饰/Kossu Ornaments

除了服装以外，靴子、围巾、帽子、团扇等服饰品也有用缂丝面料制成的。由于缂丝团扇内容较多，并在现代缂丝产品中应用较广，放在第二节中详细阐述。

In addition to clothing, boots, scarves, hats, round fans and other clothing accessories are also made of kossu fabrics. Due to many contents of kossu fabric round fan and the widely usage in modern kossu fabric products, it is elaborated in Section Two.

1. 靴子/Boots

辽代墓葬出土的缂丝凤纹女靴，上面缂织卷草纹和凤纹，图案写实、线条规整；材质上除使用了桑蚕丝以外，还加入了片金线，可谓雍容奢华，如图3-1-3所示。与辽代靴子不同，元代蒙古族靴套的图案装饰性强，具有图案化风格。

The female boots with kossu fabrics and phoenix patterns were unearthed in Liao Dynasty tombs, with kossu woven grass patterns and phoenix patterns. It possessed realistic patterns and regular lines. In addition to using mulberry silk, the material also added sliced gold thread, which could be described as graceful and luxurious（Figure 3-1-3）. Different from Liao boots, the patterns of Mongolian boots in the Yuan Dynasty were highly decorative and had a patterned style.

图3-1-3　辽代缂丝靴子
Kossu Boots in the Liao Dynasty

2. 围巾 /Scarf

辽代缂丝围巾出土于辽宁省法库县叶茂台七号墓，中间为缂丝金，两端缝制绮，设计独特、图案精美、工艺精湛，如图3-1-4所示。

The fabric scarfin the Liao Dynasty was unearthed in Tomb No.7, Yemaotai, Faku County, Liaoning Province. It had gold kossu fabrics in the middle and sewn at both ends. It had a unique design, exquisite pattern and exquisite craftsmanship（Figure 3–1–4）.

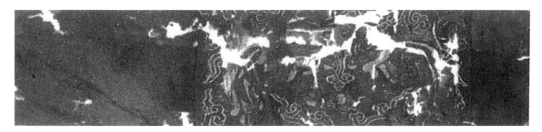

图3-1-4　辽代缂丝围巾
Fabric Scarf in the Liao Dynasty

3. 帽子/Hats

高翅帽于1974年春出土于辽宁省法库县叶茂台七号墓，由中间的圆帽和两边的高翅组成，是契丹贵族妇女戴的一种冠帽（图3-1-5）。缂丝水波地荷花摩羯纹棉帽20世纪90年代初出土于内蒙古代钦塔拉，该帽子的缂丝面料以本色丝线作为经纱，蓝、紫、褐、白丝线和金线作为纬纱进行缂织，图案以蓝色水纹为地，用金线刻出波浪，水面上漂着荷叶，鱼儿跃出水面，画面生动有趣（图3-1-6）。

A high–winged hat was unearthed in the spring of 1974 at Tomb No.7, Yemaotai, Faku County, Liaoning Province. It consisted of a round hat in the middle and high wings on both sides, and was a kind of crown hat worn by Khitan noble women（Figure 3–1–5）. The kossu–cotton hat with capricorn pattern of lotus in water weaves was unearthed in Daiqintala, Inner Mongolia in the early 1990s. The kossu fabric of this hat was woven with natural silk thread as warp yarn; with blue, purple, brown, white silk thread and gold thread as weft yarn. The base of the hat possessed a blue water pattern, and waves were carved with gold thread. Lotus leaves floated on the water surface, and fish jumped out of the water surface, which was vivid and interesting（Figure 3–1–6）.

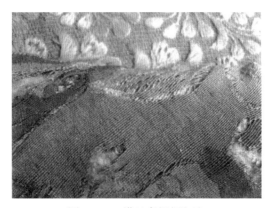

图3-1-5　缂丝高翅帽包边
Wrapping of Kossu Fabric
High-winged Hat

图3-1-6　缂丝水波地荷花摩羯纹棉帽
Kossu–Cotton Hat with Capricorn Pattern of Lotus
in Water Weaves

4. 被褥/Bedding

织造缂丝被褥所需工期长，所以主要供皇室贵族、富商大甲享用。辽宁省法库县叶茂台出土的缂金龙纹夹被，长2m，由八条宽窄和长短不同的横幅并排缝合而成。以缂丝为地，用片金线缂织海水云龙纹，每幅一升龙，姿势不同，如图3-1-7所示。

Weaving kossu fabric bedding required a long producing period, so it was exclusive for the royal family and wealthy businessmen. The golden dragon quilt unearthed in Yemaotai, Faku County, Liaoning Province, was 2 m long and was made up of eight banners with different widths and lengths stitched side by side. It took the kossu fabrics as the basis, wove the seawater cloud dragon pattern with a piece of gold thread and each dragon had different postures（Figure 3-1-7）.

图3-1-7　缂金龙纹夹被
Golden Kossu Quilt with Dragon Patterns

5. 香囊/Sachets

缂丝适合制作尺寸小、图案完整的小件饰物，如装饰袋、荷包，也就是早期的香囊。此类饰物织造时间短，又可随意装饰花纹，宋代宫廷里已经使用这种饰物。在宋代，文思院里的缂丝作和后苑造作里的缂丝作专门为帝后制作服装和日用品，也会作为外交礼物，如以缂丝药袋、缂丝篦子袋作为礼物。在清代缂丝荷包、钱袋、眼镜袋等小件饰物在宫廷中使用比较普遍，为了节省工本常常织画并用。图3-1-8所示为缂丝葫芦图案香囊，图3-1-9所示为缂丝石榴图案香囊。

Kossu fabrics are suitable for making small ornaments with small size and complete patterns, such as decorative bags and purses, which are early sachets. This kind of ornament has a short weaving time and can be decorated with patterns at will. It has been used in the court of the Song Dynasty. In the Song Dynasty, the kossu fabric in Faculty of Arts and Crafts and the kossu fabric in Houyuan specially made clothes and daily necessities for emperors, which were also used as diplomatic gifts. In the Song Dynasty, kossu fabric medicine bags and kossu fabric grate bags were given as gifts. In the Qing Dynasty, small ornaments such as kossu fabric purses, money bags and glasses bags were widely used in the court, and they were often woven and painted to save money.

Figure 3-1-8 shows the kossu sachet with the gourd pattern, and Figure 3-1-9 shows the kossu sachet with the pomegranate pattern.

图3-1-8 缂丝葫芦图案香囊
Kossu Sachet with Gourd Pattern

图3-1-9 缂丝石榴图案香囊
Kossu Sachet with Pomegranate Pattern

6. 火镰/Fire Sickle

火镰是一种取火工具，由钢条、火石和火绒组成，因其击石取火用的钢条形似镰刀而得名。火镰选用的钢条需硬度合适，淬火也需恰到好处。钢条上一般固定一个皮革、缂丝等材质的荷包，荷包内平时存放火石和火绒。火石与钢条碰击会迸发火星。火绒是用蒲绒、棉絮等易燃物沾硝或者硫黄制成，一沾火星就可点燃。

Kossu fabric fire sickle is a kind of tool for making fire, which is composed of steel bar, flint and tinder. It is named after the steel bar used for striking stone to make fire look like a sickle. The steel bars selected for fire should have proper hardness and quenching. Generally, a purse made of leather, silk and other materials is fixed on the steel bar, and flint and tinder are usually stored in the purse. Flint and steel bars will collide with mars in generating. Tinder is made of flammable materials such as cattail and cotton wool stained with nitrate or sulfur, which can be ignited when stained with mars.

图3-1-10为收藏于黑龙江省博物馆的红缂丝云蝠寿字纹荷包火镰，长10cm、宽7.5cm，由上方一如意云头状金属环、中间缂丝荷包和下方一镰刀形钢条组合而成。荷包以红色为地，中间缂金团寿纹，左右两侧以蓝、绿和白三色为主的云蝠纹成对称图案。

As shown in Figure 3-1-10, the red kossu purse with cloud-bat-longevity pattern and fire sickle collected in Heilongjiang Provincial Museum was 10 cm long and 7.5 cm wide. It was composed of a

图3-1-10 红缂丝云蝠寿字纹荷包火镰
Red Kossu Purse with Cloud-Bat-Longevity Pattern and Fire Sickle

wishful cloud-shaped metal ring at the top, a kossu fabric purse in the middle and a sickle-shaped steel bar at the bottom. The purse was made of red, with gold balls and longevity patterns in the middle, and cloud bats with blue, green and white colors on the left and right sides, which formed symmetrical patterns.

7. 扇套/Fan Cover

折扇盛行于明清时期，前后经历了两代的发展，在清代成为一种主流雅玩。从皇室贵族到平民百姓，无不以佩扇为荣。而折扇配饰也随着折扇产品增加而增加，有扇坠、扇套等。生活在古代封建制度下的人们对于信物的需求十分强烈。扇套就是用来寄托思念的众多古代手工艺品中的一种。为了进一步彰显扇子主人的身份和地位，清代发展出了形式丰富、工艺精美、质地优良的折扇扇套，制作工艺以丝织和刺绣为主，尤其是缂丝、打籽绣、盘金绣、珠绣等，为其赋予了极高的工艺价值。图3-1-11所示为清代缂丝花卉扇套。

Folding fans prevailed in the Ming and Qing Dynasties, experienced two generations of development before and after, and became a mainstream elegant stuff in the Qing Dynasty. From the royal family to the common people, they were all proud of wearing fans. Folding fan accessories also increased with the diversity of folding fan products, such as fan pendants and fan covers. People of the ancient feudal system had a strong demand for tokens. Fan cover was one of many ancient handicrafts used to show missing thoughts. In order to further demonstrate the identity and status of the fan owner, the Qing

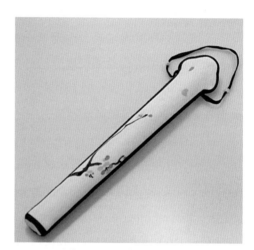

图3-1-11　缂丝花卉扇套
Kossu Fan Cover with Flower Pottern

Dynasty developed folding fan covers with abundant forms, exquisite craftsmanship and excellent texture. The production process was mainly silk weaving and embroidery, such as kossu weaving, seed embroidery, gold-plated embroidery and bead embroidery, which endowed it with extremely high technological value. Figure 3-1-11 shows the kossu fan cover with flower pattern in the Qing Dynasty.

◎ 思考题/Questions for Discussion

1. 缂丝服装主要有哪些？/What are the main types of kossu clothing?
2. 缂丝服饰品有哪些？/What are the accessories of kossu clothing?

第二节　缂丝团扇/Kossu Round Fans

一、缂丝团扇的历史沿革/Historical Evolution of Kossu Round Fans

缂丝产品具有双面效应，因此十分适合制作团扇。宋代缂丝团扇源于书画扇面，宋代缂丝团扇受到宋代绘画的影响，扇面以花鸟、山水为主，风格小巧雅致。如馆藏于台北故宫博物院的八哥桃花扇面，长25cm，宽27.8cm，扇面是花瓣形，一只八哥攀立在桃花枝上，神态专注，栩栩如生（图3-2-1）。

缂丝团扇

Kossu fabric products have double-sided effect, so they are very suitable for making round fans. The Song Dynasty kossu fabric fan originated from the painting and calligraphy sector, which was influenced by the Song Dynasty painting. The sector was mainly composed of flowers, birds and landscapes, and it had the symbol of small and elegant. For example, the kossu fan with peach blossom and myna collected in Taipei Palace Museum was 25 cm long and 27.8 cm wide, and its sector was petal-shaped. A myna climbed on the peach blossom branch, and its expression was focused and active (Figure 3-2-1).

图3-2-1　缂丝桃花八哥扇面
Kossu Fan with Peach Blossom and Myna

元代缂丝团扇比较典型的是牡丹团扇，馆藏于辽宁博物馆。扇面长22.6cm，宽26.3cm，一枝两朵相对盛开的牡丹花，一朵位于扇面中心靠右下角，继而花枝弯向扇面左上角又装饰一朵牡丹花。绿叶围绕在两朵牡丹周围，但装饰得过于繁杂，以至盖住了部分牡丹花，没有达到突出主题的效果。牡丹花用深浅不同的三种红色缂织，花叶则用孔雀绿、灰绿、黄绿三种颜色，整体色彩比较浓重。由于经丝未加捻，经丝与纬丝粗细相同，故织物的表面平滑，没有"瓦楞地"沟纹。牡丹花用木疏戗法织出花瓣深浅不同的效果，具有色彩渐渐过渡、色条规整的装饰趣味。花瓣单梭勾金，花叶、花枝边缘双梭勾金，使用金线织，大大增强了这件作品的观赏效果（图3-2-2）。

The typical kossu round fan in the Yuan Dynasty was the peony round fan, which was collected in Liaoning Museum. The fan was 22.6 cm long and 26.3 cm wide. There were two peony flowers in full bloom, one was located in the lower right corner of the fan, and then the flower branch bent to the upper left corner of the fan and was decorated with another peony flower.

图 3-2-2 缂丝牡丹团扇
Peony Kossu Round Fan

Green leaves surrounded the two peonies, but the decoration was too complicated so that the peony flowers were covered and it failed to achieve the effect of highlighting the theme. Peony flowers were woven in three kinds of red with different shades, while flowers and leaves were made of malachite green, gray-green and yellow-green, the color was heavy generally. Because the warp yarn was not twisted and the thickness of warp yarn and weft yarn was the same, the surface of the fabric was smooth and there was no "corrugated ground" groove. Peony flowers were woven with different shades of petals by the wood sparse method, which had the decorative interest of gradual color transition and regular color strips. The petals were hooked with gold by a single shuttle, and the edges of flowers, leaves and branches were hooked with gold thread, which greatly enhanced the ornamental effect of this work（Figure 3-2-2）.

明代缂丝秋蟾团扇，也是历代缂丝团扇扇面中非常罕见的作品，此扇目前为故宫博物院调拨至新疆博物馆的文物，长24.6cm，宽23.5cm。此图案缂织的是形象生动的秋天景象、石树葱秀，毫素间有。图案疏密有致，构图紧凑、一只蟾从湖中爬上岸边，被刻画的形态似乎非常警觉，用精练的技艺织出蟾爬行之状和花叶萧疏的清秋气氛。蟾代表月亮，秋色夜中景，月中有蟾，故改月为蟾，蕴含着多种象征和图腾的意义，此团扇构图体现了明代工笔花鸟画的许多基本特色。使用了单双子母经，使断纬和经线的结合更加牢固，这是前代缂丝中罕见的技法，图案及品艺高雅，是研究织物技法与绘画艺术风格的珍贵实物（图3-2-3）。

The autumn toad kossu round fan in the Ming Dynasty was also a very rare work among the kossu fabric round fans in past dynasties. This fan is a cultural relic transferred from the Palace Museum to Xinjiang Museum, with a length of 24.6 cm and a width of 23.5 cm. This pattern weaved a vivid autumn scene, with beautiful stone and trees. The pattern was dense and the composition was compact. A toad climbed up the shore from the lake, and it seemed

图 3-2-3 缂丝秋蟾团扇
Autumn Toad Kossu Round Fan

to be very alert. The crawling shape of a toad and the clear autumn atmosphere with sparse flowers and leaves were depicted with refined skills. Toad represented the moon, the autumn night scene, and there were toads in the moon, so the moon was changed to the toad, which contained a variety of symbolic and totem meanings. The composition of this round fan embodied many basic characteristics of meticulous flower–and–bird painting in the Ming Dynasty. The use of single and twin Zi Mu Warps weaving made the combination of broken weft and warp threads firmer. This was a rare technique in the previous generation of silk reeling, with elegant patterns and art, and was a precious object for studying fabric techniques and painting artistic styles（Figure 3–2–3）.

　　清代缂丝团扇题材丰富，目前收集的缂丝团扇实物资料主要为动植物纹样。其中动植物题材扇面多与花卉纹样组合，多花鸟图、花蝶图等，纹样寓意吉祥；图案兼具写实、装饰风格。如故宫博物院馆藏的红色缂丝海屋添筹图面乌木雕花柄团扇，该扇通柄长46cm、面径34.3cm，画面以红色、黄色为主，宫殿般的海屋辅以海水、江崖、灵芝，寓意长寿。又如红色缂丝兰花图面象牙八仙图柄团扇，扇作上广下狭的芭蕉式，扇面中央于桃红色地上缂织折枝牡丹、梅花及绶带鸟，其铺排全依画理，花枝偃仰有致，小鸟尤其生动。值得注意的是，物象形体边缘以丝线缂出，其内则为添笔彩绘，这样既降低了工艺难度，又使局部更为精致和写实，反映出清代缂丝的新风格（图3–2–4，图3–2–5）。

The Qing Dynasty kossu fabric round fan is rich in theme. According to the material data of current collection of kossu fabric round fan, animal and plant patterns are mainly collected. Among them, the fan of animal and plant themes was mostly combined with flower patterns, such as flower and bird pictures and flower butterfly pictures, which symbolized auspiciousness; the pattern was both realistic and decorative. For example, the red kossu fabric cabin in the Palace Museum added a black wood carving flower handle round fan, which was a kossu fabric round fan with a handle length of 46 cm and a face diameter of 34.3 cm. The picture was mainly red and yellow. The palace–like cabin was supplemented by seawater, river cliffs and Ganoderma lucidum, which symbolized prolonged life. Another example was the red kossu round fan with orchid design and eight–immortal ivory handle, which was a Basho style with a wide top and narrow bottom. In the center of the fan, folded peony, plum blossom and ribbon birds were woven on the pink base. Its layout followed the painting principle, and the flowers were special, especially the birds. It was worth noting that the edge of the object shape was cut out with silk thread, and the inside was painted with a pen, which not only reduced the difficulty of the process but also made certain part more delicate and realistic, reflecting the new style of kossu fabric in the Qing Dynasty（Figures 3–2–4 and 3–2–5）.

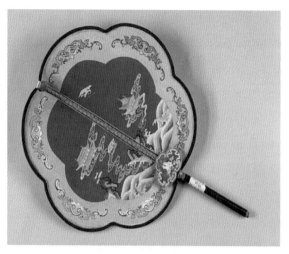

图3-2-4　红色缂丝海屋添筹图面乌木雕花柄团扇
Red Kossu Round fan with Blessing Longevity Design
and Carred Flower Black Wood Handle

图3-2-5　红色缂丝兰花图面象牙八仙图柄团扇
Red Kossu Round Fan with Orchid Design and
Eight-immortal Ivory Handle

二、缂丝团扇的材质/Materials of Kossu Fabric Round Fans

（一）扇面/Fan Face

团扇也称纨扇（纨，系古代一种很细的绢），由于扇面形如满月，故名团扇；因宫廷大量使用，又称宫扇。缂丝团扇的扇面主要为桑蚕丝线（包括生丝线和熟丝线），除此以外还有金线等。

The round fan is also called Wan Fan（Wan, very thin silk in ancient times）, which is named round fan because its fan is shaped like a full moon. Because it was widely used in the court, it was also called court Fan. The fan surface of the kossu fabric round fan is mainly mulberry silk thread（including raw silk thread and gloss silk thread）, in addition to the gold thread, etc.

（二）扇柄（扇骨）/Fan Handle（Fan Bone）

缂丝团扇以缂丝面料为扇面，木、竹、骨等材为执柄；团扇的扇柄有木柄、红木柄、棕竹柄、留青竹柄、斑竹柄、玉柄、象牙柄、骨柄、紫漆柄、蓝漆柄、金漆柄、黑漆柄等。

The kossu fabric round fan takes kossu fabric as the fan surface; and wood, bamboo, bone and other materials as the handle. The fan handle of the round fan has wooden handle, mahogany handle, brown bamboo handle, green bamboo handle, mottled bamboo handle, jade handle, ivory handle, bone handle, purple lacquer handle, blue lacquer handle, gold lacquer handle, black lacquer handle and so on.

古人不仅注重扇柄材料的选用，而且十分重视扇柄的雕刻工艺。扇柄（扇骨）分"素骨"和"有工"两类。"素骨"是不加任何装饰和雕刻的纯竹木类扇骨，呈现的是竹木天然的机理，体现的是自然的美感。"有工"是施以工艺的扇骨，采用雕刻、镶贴、镶嵌、漆画、镂空、透雕、烫花、手绘等方式对扇骨进行装饰，其图案丰富多彩、惟妙惟肖，体现了传统工艺的价值。另有玉、翠、犀角、象牙、宝石种种材料点缀，再佩之扇坠、流苏装饰。

The ancients not only paid attention to the selection of fan handle materials but also attached great importance to the carving process of fan handle. Fan handle (fan bone) is divided into two categories: "plain bone" and "Yougong". "Plain bone" is a pure bamboo and wood fan bone without any decoration and carving, which presents the natural mechanism of bamboo and wood and embodies the natural aesthetic feeling. "Yougong" is a fan bone with craftsmanship, which is decorated by carving, inlaying, lacquer painting, hollowing out, transparent carving, ironing and hand painting. Its patterns are colorful and vivid, which embodies the value of traditional crafts. There are also jade, emerald, rhinoceros horn, ivory and precious stones, which are decorated with fan pendants and tassels.

在团扇的制作中，扇面和扇柄的结合有很多种形式，一种是用金属丝加工成扇缘形状，将扇面蒙绷于扇缘之上，再将扇骨贴合于扇面之上；另一种是将金属丝扇缘的结合点合并起来，穿插固定在扇骨之中；还有一种是将扇骨劈开，夹合扇面等。但其共同点都是扇面紧绷，形成展开的平面，使其能够继续进行绘画、贴镶等装饰。雕刻艺术与书画艺术的完美结合，使得中国扇子成为许多文人雅士喜爱和渴求的艺术藏品，也成为展现中国独特文化魅力的艺术结晶。

In the production of round fans, there are many forms of combination of fan face and fan handle. One is to process the sector into the shape of an edge with metal wire, cover the sector over the fan edge, and then attach the fan bone to the sector; the other is to combine the joint points of the fan edges of metal wires and fix them interspersed in the fan bones; another is to split the fan bone and clamp the fan surface. But their common point is that the sector is tight, forming an unfolded plane, which enables them to continue painting, sticking and doing other decorations. The perfect combination of carving art and painting and calligraphy art makes Chinese fans a favorite and eager art collection of many literati, and also an artistic crystallization showing the unique cultural charm of China.

三、缂丝团扇的题材与形制/Themes and Shapes of Kossu Round Fans

（一）缂丝团扇的题材/Themes of the Kossu Round Fans

缂丝团扇内容以鸟兽虫鱼、园林山水、花草树木、人物肖像为主。亦有团扇书法，书画作为扇子的装饰艺术存在，也借扇子千姿百态的装饰风格，形成中国书画的特殊构图形式。

The content of kossu round fans is mainly birds, animals, insects and fish, gardens and landscapes, flowers and trees, and individual portraits. There are also round fan calligraphy, painting and calligraphy as the decorative art of fans, and the special composition form of Chinese painting and calligraphy is formed by the various decorative styles of fans.

（二）缂丝团扇的形制/Shapes of the Kossu Round Fans

缂丝团扇具有展示双面效应的优势，缂丝团扇自元代出现到清代时已经制作得非常精美。缂丝团扇至今，更多的功能不是纳凉，而是彰显文人情怀，成为践行传统工艺精神的载体。从形制上看，团扇不仅仅是圆形，还出现了方形、椭圆形、六角形、八角形、瓜楞形、花瓣形、桐叶式、芭蕉叶式，以及其他异形。其中海棠形、梅花形是除圆形之外较常见的（图3-2-6~图3-2-10）。

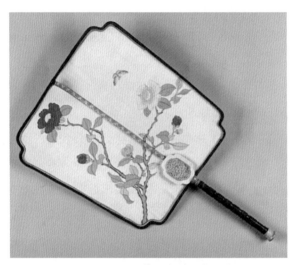

图3-2-6 委角长方形团扇
Rectangular Round Fan with Cutoff Corners

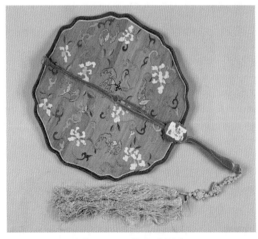

图3-2-7 葵花式团扇
Sunflower Shaped Round Fan

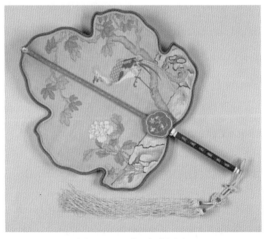

图3-2-8 桐叶式团扇
Tung-leaf-shaped Round Fan

Kossu fabric round fan has the advantage of displaying a double-sided effect, and it had been beautifully made since the Yuan Dynasty to the Qing Dynasty. Up to now, the function of the kossu fabric round fan is not only to enjoy the cool, but to show the literati's feelings and become the carrier of practicing the traditional craft spirit. From the shape point of view, the round fan is not only round but also square, oval, hexagonal, octagonal, melon-shaped, petal-shaped, tung-leaf-shaped, banana-leaf-shaped and other special-shaped. Among them, begonia-shapeds and plum-blossom-shaped are more common except for round shapes (Figures 3-2-6 to 3-2-10).

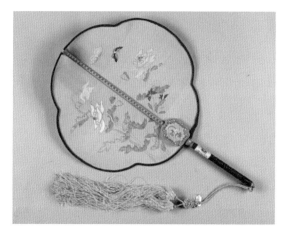

图3-2-9　海棠式团扇
Begonia-shaped Round Fan

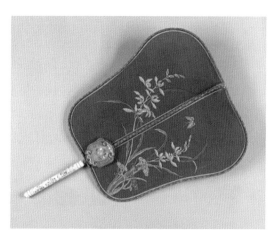

图3-2-10　芭蕉叶式团扇
Banana-leaf-shaped Round Fan

◎ 思考题/Questions for Discussion

1. 缂丝团扇的扇面题材有哪些？/What are the fan themes of kossu fabric round fan?

2. 简述缂丝团扇扇骨的分类。/Introduce the classification of the fan bones of the kossu round fan.

3. 缂丝团扇的形制有哪些？/What are the shapes of kossu round fan?

第三节　缂丝宗教用品/Kossu Religious Supplies

两千多年前，我们的先人在丝绸之路回途中带回了西亚的梭织技术，同时在邻国印度，有一种先进的哲学思想正在形成，那就是佛教。早在唐代，就有用缂丝制品装点佛经，制作僧人的袈裟等。唐代时佛教盛行，从目前出土的缂丝产品看，唐代缂丝主要用于宗教用品中。具体的有敦煌出土的浅橙地花卉纹缂丝带、蓝地十样花缂丝带、小花缂丝经卷系带等。

缂丝宗教用品

More than 2,000 years ago, our ancestors brought back the weaving technology of West Asia on the way of the Silk Road. At the same time, in our neighboring country India, an advanced philosophical thought was formed referring to Buddhism. As early as the Tang Dynasty, kossu fabric products were used to decorate Buddhist scriptures and make monks' cassocks, with Buddhism being prevailing, the kossu fabric in the Tang Dynasty was mainly used in religious articles from the products unearthed at present. Specifically, there are the light orange flower pattern kossu ribbon, ten kinds of flowers on the blue base kossu ribbon, small flowers sutra kossu ribbon unearthed in Dunhuang and so on.

一、唐代宗教用品/Religious Supplies in the Tang Dynasty

1. 浅橙地花卉纹缂丝带/Light Orange Flower Pattern Kossu Ribbon

浅橙地花卉纹缂丝带出土于敦煌，目前由大英博物馆馆藏，共两条，其中一条丝带长16.6cm、宽1.4cm，另一条丝带长9.4cm、宽1.3cm。纬线为浅橙色作地，白、绿、蓝、橙、紫等显花，花卉纹样，可能为宝花图案的局部，如图3-3-1所示。

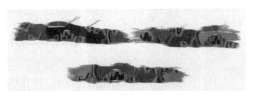

图3-3-1 浅橙地花卉纹缂丝带
The Light Orange Flower Pattern Kossu Ribbon

There are two light orange flower pattern kossu ribbons unearthed in Dunhuang, which are currently collected by the British Museum. one ribbon was 16.6 cm long and 1.4 cm wide, and the other ribbon was 9.4 cm long and 1.3 cm wide.The weft lines were based on light orange and white, green, blue, orange and purple flowers patterns, which may be part of the precious flower pattern, as shown in Figure 3-3-1.

2. 蓝地十样花缂丝带/Ten Kinds of Flowers on the Blue Base Kossu Ribbon

蓝地十样花缂丝带长18.5cm、宽1.5cm；经线为多根丝线以S捻并合、单根排列浅米色丝线，密度为18根/cm；纬线为蓝、棕、白、黄等色丝线及片金线，单根排列，密度约72根/cm；狭长形缂丝带，在蓝色地上以各色丝线缂织出十样花卉纹样，图案经向循环为5.7cm，其风格与浅橙地花卉纹缂丝十分接近，做工极为精致。但为了增强图案的装饰效果，此件缂丝在局部使用了片金线，从放大照片来看，这种片金线的背后采用了纸质背衬。这是目前所知最早的纸质背衬的片金实物，如图3-3-2所示。据斯坦因推测，这件作品主要用于缂丝悬绊。

This kind of ribbon was 18.5 cm long and 1.5 cm wide. The warp threads were multi-threads twisted and merged in the shape of S, and the single one was lined with light beige threads with a density of 18 threads/cm. The weft threads were blue, brown, white and yellow silk threads and gold threads, which were arranged in a single thread with a density of about 72 threads/cm. Ten flower patterns were woven with various silk threads forming long and narrow kossu fabric ribbons on the blue base, and the warp cycle of the patterns was 5.7 cm. Its style was very close to that of the light orange flower pattern kossu fabric, and its workmanship was extremely exquisite. However, in order to enhance the decorative effect of the pattern, flake gold thread was used locally in this piece of kossu fabric. From the enlarged photos, a backing paper was used behind this flake gold thread. This was the earliest known flake gold thread with paper backing, as shown in Figure 3-3-2. According to Stein's speculation, this work was mainly used for kossu hanging.

图3-3-2 蓝地十样花缂丝带及细节图
The Ten Kinds of Flowers on the Blue Base Kossu Ribbon and Its Details

3. 小花缂丝经卷系带/Small Flowers Sutra Kossu Ribbon

小花缂丝经卷系带出土于敦煌，目前收藏在法国国家图书馆，该系带系缚《阿弥陀经》，这是敦煌藏经洞中唯一的用缂丝带作经卷系带的例子。带很窄，宽仅0.7~0.8cm，共有13根加有S捻的白色丝线。长度方向已残，为24.4cm，其纬线的色彩保存相当完好，蓝色作地，红、粉红、绿、草绿、草黄、白、湖蓝等多种色彩显花，花型很小，为六角形的主花和两个三角形的宾花相间排列，主花由绿和湖蓝两种色彩交替，宾花由红和白交替，在极小的空间中，作了尽可能多的变化，体现了缂丝艺术的特点，如图3-3-3所示。藏经洞中所出缂丝带亦有多例，主要用作幡首的系带、经帙上的装饰带，风格基本一致，但以此件最为简洁。

（a）细节图/Detail Drawing

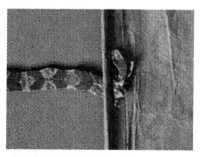

（b）缂丝系在经卷上/The Silk Is Tied to the Scroll

图3-3-3 小花缂丝经卷系带
The Small Flowers Sutra Kossu Ribbon

The small flowers sutra kossu ribbon was unearthed in Dunhuang and is currently collected in the National Library of France. This ribbon was bound with *Amitabha*

Sutra, which was the only example of using kossu ribbons as sutra ribbons in Dunhuang Sutra Cave. The belt was very narrow, only 0.7 ~ 0.8 cm wide, and there were 13 white silk threads with twists by "S" and it was 24.4 cm long. The color of its weft was well preserved, which blue was the base, and red, pink, green, grass green, grass yellow, white, pale blue and other colors were colored for small flowers pattern. The hexagonal main flower and two triangular sub–flowers were arranged alternately. The main flower was alternated by green and pale blue, and the sub–flower was alternated by red and white. In a very small space, it had been changed as much as possible, embodying the art of kossu fabric as shown in Figure 3–3–3. There were also many cases of kossu ribbons produced in the Sutra Cave, which were mainly used as the lace of steamer and the decorative ribbons on a folder file of volume. The style was similar, but this one was the most concise.

4. 绢地彩绘幡头/Silk Drawing Streamer

幡的残片，如图3-3-4所示，共三片，其中一块为幡正面，另外两块褐色绢织物则均作衬里之用。该幡的幡面及幡身部分由同一片本色绢制成，三角形幡面高10.2cm，宽16.1cm，以蓝色作地，中心绘一大朵花卉纹样，三角则各装饰有一朵小花；残留的幡身高约4.2cm，以红、橙两种颜色为主色绘出直条形的帷幔图案，幡面与幡身间绘一条墨线以示区别；幡头两侧以宽2.7cm的红地团窠立鸟纹缂丝织物包边，团窠高4.3cm，宽2.7cm，图案中心为一立鸟纹样，四周装饰有四片绿色花瓣及四朵花蕾，花瓣中心及花蕾边缘以片金线缂织而成，团窠采用二二错排，上下两团窠中的立鸟朝向不同，不同团窠中立鸟及其背景颜色常有变化。在幡头上端钉有一长约4.5cm的橙红色绢制祥扣。

(a) 缂丝部分/Part of Kossu Fabric

(b) 完整图/Complete Drawing

图3-3-4 绢地彩绘幡头
Silk Drawing Streamer

Fragments of the streamers were consisted of three pieces, as shown in Figure 3–3–4, one of which was the front of the streamer, and the other two pieces of brown silk fabric were used for lining. The surface and body of the streamer were made of the same piece of natural silk. The triangular streamer surface was 10.2 cm high and 16.1 cm wide. It was made to use blue as the base, and a large flower pattern was painted in the center and a small flower was decorated in each triangle. The height of the remaining streamers was about 4.2 cm, and a drapery pattern in a vertical element was drawn with red and orange as the main colors, and an ink line was drawn between the streamer surface and the streamer body to show the difference. The two sides of the streamer head were wrapped with the vertical bird in the nested pattern kossu fabric with a height of 4.3 cm and a width of 2.7 cm. The pattern in the center was a vertical bird pattern, which was decorated with four green petals and four flower buds. The center of the petals and the edge of the flower buds were woven with flake gold thread. The nest was arranged in two staggered rows with the vertical birds in the upper and lower nests having different directions. The vertical birds, as well as their background colors, often changd in different nests. An orange–red silk loop with a length of about 4.5 cm was nailed at the upper end of the streamer head.

二、西夏宗教用品 / Religious Supplies in the Kingdom of Xia

西夏又被称作"佛国"，举国信仰佛教，所以相比较而言对于佛教纹样的应用也是居于辽金之上的。例如，西夏墓出土的两件缂丝唐卡，其佛教人物纹样十分精美。

The Kingdom of Xia was also called a "Buddhist country", and the whole country embraced Buddhism, so the application of Buddhist patterns was also more than the Liao and Jin Dynasties, For example, two pieces of kossu fabric Thang–ga unearthed from the Western Xia Tomb, they embraced the exquisited patterns of Buddhist figures.

三、元代宗教用品 /Religious Supplies in the Yuan Dynasty

元代宗教用缂丝数量明显增加，以佛教和道教题材的作品为主。元代盛行佛教，将作院设立织佛像提举司，专门织造宫廷佛像，其中除了织锦外，还应有缂丝。另外，唐卡、经袱用缂丝、织锦等珍贵的材料织成。从道教演化而来的祝寿题材的作品，如《八仙祝寿》《东方朔偷桃》等内容造的缂丝作为雅俗共赏的礼物受到各个阶层人士的青睐。佛经、佛具的盖幅也用缂丝材料制作。1955年北京西长安街双塔庆寿寺出土的缂丝《莲塘鹅戏图》，长68cm，宽56cm，紫色底，上面有黄绿相间的水波纹和卧莲，卧莲之间有鹅浮游其中。缂丝佛经盖幅流传稀少，现存多为明代作品。

In the Yuan Dynasty, the number of religious kossu fabrics increased obviously, taking Buddhist and Taoist works as the principal theme. Buddhism prevailed in the Yuan Dynasty, and a Buddha weaving and lifting department was set up in the institute to weave Buddha statues, which should be applied kossu fabric besides brocade to these statues. In addition, Thang–ga and Confucian classics package was woven with precious materials such as kossu fabric and brocade. Birthday works evolved from Taoism, such as *Eight Immortals to Celebrate Birthday* and *Dong Fangshuo Stealing Peach*, are favored by people from all walks of life as gifts that enjoy both refined and popular tastes. The covers of Buddhist scriptures and Buddhist utensils were also made of silk. In 1955, *the Picture of Gannet in a Lotus Pond* with kossu fabric was unearthed from twin towers of Qingshou Temple, West Chang'an Street, Beijing. It was 68 cm long and 56 cm wide, on the purple ground, with yellow and green water ripples and lying lotus which float geese.The cover of kossu Buddhist scriptures is rare, and there are still the Ming Dynasty works.

缂丝《东方朔偷桃》，如图 3-3-5 所示，长 58.5cm，宽 33.5cm，馆藏于故宫博物院。此图题材取自东方朔偷桃的典故。画面表现东方朔从仙界偷桃后疾走之状，人物的胡须和飘曳的衣裾，显出疾走的动态。东方朔手持偷摘的蟠桃，回首环顾，面露窃喜，其偷桃得手后的得意之情和担心被仙吏发现的微妙心理被刻画得惟妙惟肖。画面上方祥云缭绕中累累仙桃悬垂枝头，下方配以灵芝、水仙和竹石，以谐音寓意"芝仙祝寿"。画面钤"乾隆御览之宝""乾隆鉴赏""三希堂精鉴玺""宜子孙"和"秘殿珠林"。此幅缂丝图以蓝和浅蓝为主调，配有石青、月白、瓦灰等色。运用平缂、木梳戗等缂织技法。以平缂作色块平涂，在纹样边缘或二色相遇处则使用勾缂进行勾勒，或以长短戗进行调色过渡。寿石用深蓝、蓝和浅蓝三晕色戗缂，以突出山石的立体感。尤以合色线技法颇有特色，如东方朔的手指缝用黑、白二色丝；灵芝的茎部用石青和米色丝表现茎干的糙涩感。整幅作品设色简洁，气韵生动。

Dong Fangshuo Steals Peach in kossu fabrics, as shown in Figure 3-3-5, with a length of 58.5 cm and a width of 33.5 cm, is collected in the Palace Museum. The theme of this picture was taken from the allusions of Dong Fangshuo stealing peaches. The picture showed Dong Fangshuo scurrying after stealing peaches from the celestial world, revealing the state from the beards and

floating clothes of the characters. Dong Fangshuo held the stolen flat peach, looked back and showed secretly pleased. His pride after stealing the peach and his subtle psychology of worrying about being discovered by fairy officials were vividly depicted. At the top of the picture, there were many branches of peaches in the auspicious clouds, and at the bottom, there were Ganoderma lucidum, daffodils and bamboo stones, which symbolized "birthday congratulation and Longevity" with homophonic meaning. The pictures showed "Treasure of Qianlong Imperial View" "Appreciation of Qianlong" "Fine Seal of Sanxi Hall" "Appropriate for Descendants" and "Pearl Forest of Secret Hall". This kossu

图 3-3-5　缂丝《东方朔偷桃》
Dong Fangshuo Stealing Peach in Kossu Fabrics

picture was mainly blue and light blue, with azurite, moon white, tile gray and other colors. Weaving techniques such as flat weaving and Qiang weaving were used. The picture regarded flat weaving as color block, which was drawn by the outline at the edge of the pattern or where the two colors met, or color matching was applied to transit with long–and–short "draw weaving". The stones used dark blue, blue and light blue to highlight the three–dimensional sense of rocks. In particular, the technique of combining color lines was quite distinctive, such as Dong Fangshuo's finger sewing with black and white silk; the stem of Ganoderma lucidum used azurite and beige silk to express the roughness of the stem. The whole work was simple in color and vivid in flavor and tone.

四、清代宗教用品 /Religious Supplies in the Qing Dynasty

清代的宗教主要是佛教。从努尔哈赤开始尊崇佛教，后来的皇太极、顺治、康熙和乾隆皇帝也都尊崇佛教。道教是传统的宗教，在民间广泛传播，并日益世俗化，因此清政府对道教持宽容的态度。为了参加某些宗教仪式之需要，皇室人员也制作宗教服装，缂丝可谓是其中奢侈的面料。其中袈裟和道袍的数量比较多。香港贺祈思收藏基金会藏康熙缂丝袈裟，身长111cm，通袖宽291cm。袈裟上的"百纳补"构图是赤贫的象征，也是纪念佛祖常穿以碎片缝补而成的衣服，以示看破荣华富贵。背景是黄色，上面有龙纹和凤，说明这件袈裟与皇室有关。

The religion of the Qing Dynasty was mainly Buddhism. Buddhism was respected by the founding emperor Nurhachi, and the later emperors Huang Taiji, Shunzhi, Kangxi and Qianlong also respected Buddhism. Taoism was a traditional religion, which spread widely among the people and became increasingly secular, so the Qing government held a tolerant attitude towards Taoism. In order to attend some religious ceremonies, the royal family also made religious costumes, and kossu fabric was the luxurious material. Among them, the number of cassocks and Taoist robes was relatively large.The Kangxi silk cassock collected by Hong Kong He Qisi Collection Foundation was 111 cm long and 291 cm wide. The composition of "Bainabu" on the cassock was a symbol of extreme poverty, and it also often wore clothes sewn with fragments to commemorate Buddha, showing acknowledge many on splendor and wealth. The background was yellow, with a dragon pattern and phoenix on it, which showed that this cassock was related to the royal family.

1. 道袍/Taoist Robes

该道袍馆藏于美国大都会博物馆。这件耀眼的衣服本来是道士在仪式上穿的，此处展示的长袍背面描绘了五条龙盘旋在从海中升起山脉的景观之上，如图3-3-6所示。

This Taoist robe is collected in the Metropolitan Museum of America. This dazzling dress was

图 3-3-6 道袍
A Taoist Robe

originally worn by Taoist priests at the ceremony. The back of the robe shown here depicted five dragons hovering over the landscape of mountains rising from the sea, as shown in Figure 3-3-6.

2. 缂丝密集金刚像/The Statue of Guhyasamaja in Koosu Fabrics

缂丝密集金刚像，如图 3-3-7 所示，清乾隆时期，长 100cm，宽 74cm。此作品是苏州织匠严格按照西藏佛像画稿为皇宫织做的。像背后有汉、满、蒙、藏四体文字款识："乾隆四十六年十一月初五日，钦命章嘉胡土克图认看供奉'利益阳体秘密佛'（密集金刚）。"密集金刚身蓝黑色，三面，正中蓝黑色，右面白色，左面红色。六臂，胸前两手交叉持金刚铃、杵，右二手持法轮、莲花，左二手持摩尼宝、短剑。胸前拥抱浅蓝色可触金刚佛母，亦为三面六臂，持物与主尊相同。金刚浅绿色头光，橘红色身光，套以红黄蓝白四色光环。天空祥云正中坐不动佛，两边是四位佛教祖师。下界是两位护法神。

The Statue of Guhyasamaja in koosu fabric, as shown in Figure 3-3-7, had a length of 100 cm and a width of 74 cm from the Qianlong period in the Qing Dynasty. This work was woven by Suzhou weavers for the Imperial Palace in strict accordance with the Tibetan Buddha statues. There were four characters in Han, Manchu, Mongolian and Tibetan behind the image: "On the fifth day of November in the forty-sixth year of Qianlong, Zhang Jia Hutuketu was appointed to recognize and consecrate the secret Buddha of the 'benefit positive body' (Guhyasamaja)." The body of Guhyasamaja was blue-black with three sides, blue-black in the middle, white on the right side and red on the left side. He had six arms, crossed hands in front of the chest holding diamond bells and pestles, and wheel as well as lotus flowers on the right, in addition, Cintamani and

图 3-3-7 缂丝密集金刚像
The Statue of Guhyasamaja in Kossu Fabrics

dagger on the left, embracing the light blue Buddha mother in front of his chest, she was also three sides and six arms, and her holding objects were the same as that of the Lord. Buddha's warrior attendant wore a light green headlight and orange body light, setting with a red, yellow, blue and white aura. The Mitukpa sat in the middle of the auspicious clouds in the sky. There were four Buddhist ancestors on both sides. What's more, two guardian gods were below them.

3. 缂丝加绣观音像轴/The Statue Shaft of Guanyin in Kossu Fabrics and Embroidery

缂丝加绣观音像轴，如图3-3-8所示，清乾隆时期，长147cm，宽60cm，馆藏于故宫博物院。此像描绘观音身着珠宝璎珞装饰的天衣彩裙，立于五彩祥云中的莲台之上，身后背光辉映，头上华盖笼罩，供拜礼敬阿弥陀佛。观音身两侧各21只手，或持法器，或结手印，为汉地千手观音形象。观音又身披人兽，此为藏传佛教影响之痕迹。画幅运用齐缂、缂金、构缂等技法缂织人物及其衣饰，观音之披帛则用缂线钉绣，表现出轻纱的质感和透明感，同时在某些细部以敷彩、敷金等绘画手法表现出缂、绣难以达到的效果。多种艺术手法的综合运用使观音之艺术形象的塑造更为丰满完美，堪称乾隆时期缂丝加绣艺术的优秀代表作。钤"秘殿珠林""秘殿新编""珠林重定""三希堂精鉴玺""宜子孙""乾清宫鉴藏宝""太上皇帝之宝""乾隆御览之宝""乾隆鉴赏""宣统御览之宝"等十玺。

图3-3-8 缂丝加绣观音像轴
The Statue Shaft of Guanyin in Kossu Fabrics and Embroidery

The statue shaft of Guanyin in kossu fabrics and embroidery from Qianlong in the Qing Dynasty is shown in Figure 3-3-8, with a length of 147 cm and a width of 60 cm, collected from the Palace Museum. This image depicted Guanyin dressed in a colorful dress decorated with jewelry and jewelry of precious stones, standing on a lotus platform in colorful clouds, with golden light behind her and a canopy over her head for worshipping Amitabha. There were 21 hands on each side of Guanyin's body, holding instruments or palms together, which was the image of Guanyin with thousands of hands in the Han Dynasty. Guanyin was also covered with people and animals, which was a trace of the influence of Tibetan Buddhism. The figures and their clothes were woven by techniques such as Qike technique, gold thread and structure weaving, while the silk shawl of Guanyin was embroidered with thread nails in topstitching, showing the texture and transparency of fine gauze. At the same time, in some details, painting techniques such as applying color and gold showed the effect that was difficult to achieve by weaving and embroidery. The comprehensive application of various artistic techniques made the artistic image of Guanyin fuller and more perfect, which can be called an excellent representative work of kossu fabric and embroidery art in the Qianlong period. Ten seals, such as "Pearl Forest of Secret Hall" "New Edition of Secret Hall" "Re-determination of Pearl Forest" "Fine Seal of Sanxi Hall" "Appreciation of descendants" "Qianqing Palace Treasures" "The Treasure of Emperor's father" "Treasure of Qianlong Imperial View" "Appreciation of Qianlong" and "Xuantong Imperial View Treasure", etc.

4. 缂丝乾隆御笔心经册页/The Album of Heart Sutra Wrote by Qianlong in Kossu Fabrics

缂丝乾隆御笔心经册页，如图3-3-9所示，清乾隆时期，八开，每开长19cm，宽18cm，馆藏于故宫博物院。此册页摹缂乾隆帝于乾隆十八年（1753年）元旦所书《般若波罗蜜多心经》。作品在粉色地上，墨线缂织经文。前后副页彩缂观音大士和韦陀像以及彩云金龙等纹样。采用平缂、搭缂和构缂等技法缂织。心经字迹结体圆润，人物及云龙形象亦较生动。册页首尾缂"写心""乾隆宸翰""荑荛堂"，钤"得大自在"。

The Album of Heart Sutra was written by Qianlong in kossu fabrics from Qianlong in the Qing Dynasty in eighty percent, as shown in Figure 3–3–9. Each ten percent was 19 cm long and 18 cm wide, collected by the Palace Museum. This album copied the Prajna Polomi Heart Sutra written by Emperor Qianlong on New Year's Day in the 18th year of Qianlong (1753). The work was on the pink base, and the scriptures were woven with ink thread. The front and back sub-pages were decorated with patterns such as Guanyin and Wei Tuo, as well as colorful clouds and golden dragons. It also used techniques such as flat weaving, shuttling weaving and structure weaving. The handwriting of the Heart Sutra was mellow, and the characters and cloud–dragon images were vivid. At the beginning and end of the album, it weaved "Writing Experience" "Square Relief" and "Ti Rao hall" and sealed "Be at ease".

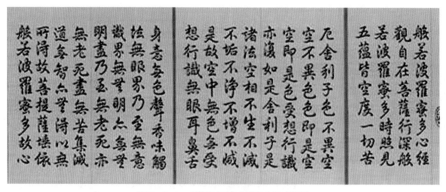

图3-3-9　缂丝乾隆御笔心经册页
The Album of Heart Sutra Wrote by Qianlong in Kossu Fabrics

五、现代宗教用品/Modern Religious Supplies

现代宗教缂丝保留清代缂丝宗教用品的种类，主要为缂丝佛像、唐卡等，如《观无量寿经》，如图3-3-10所示。《观无量寿经》由苏州祯彩堂整理，参照经书所述，花了3年时间制作缂丝用绘稿白描和彩稿。祯彩堂在2003年、2012年和2015年（完成时间）曾经做过三次不同版本的复制。此幅为2012年完成，宽2m，长2m。参与绘制、配色、造机等制作全过程的人员有10余人。作品有佛像人物581名，楼阁莲池、宝幢、飞天密布整个画面，内容庞大，各区各段有经书要义旁白，是缂丝艺术表现完整佛经的有依有据的作品。

The types of religious stuff reserved by modern religious kossu fabric in the Qing Dynasty were mainly silk Buddha statues, Thang-ga, etc., such as *Amitayurdhyana Sutra*, as shown in Figure 3-3-10. *Amitayurdhyana Sutra* was sorted out by Zhen Cai Tang in Suzhou, referring to the scriptures, and it took three years to make the kossu fabric with painted manuscripts and colorful manuscripts. Zhen Cai Tang made three copies of different versions in 2003, 2012 and 2015 (completion time). This picture was completed in 2012, with a width of 2 meters and a length of 2 meters. More than 10 people participated in the whole process of drawing, color matching, machine making and production. There were 581 Buddha figures in the works, including pavilions, lotus ponds, and treasure buildings and flying apsaras, which were densely covered with the whole picture and had huge contents. Each section in each part had narration and essence of scriptures. Thus it is a well-founded work that expresses the complete Buddhist scriptures in the art of kossu fabric.

图3-3-10 缂丝《观无量寿经》
Kossu *Amitayurdhyana Sutra*

◎ **思考题/Questions for Discussion**

列举缂丝宗教用品的代表作。/List some masterpieces of Kossu Fabrics religions supplies.

第四节　缂丝其他产品/Other Kossu Products

一、缂丝屏类/Screen Class in Kossu Fabrics

1. 缂丝屏风/Screen in Kossu Fabrics

缂丝其他产品

屏风是中国传统建筑物内部挡风用的一种家具，所谓"屏其风也"。屏风作为传统家具的重要组成部分，历史由来已久。屏风一般陈设于室内的显著位置，起到分隔、美化、挡风、协调等作用。它与古典家具相互辉映，相得益彰，浑然一体，呈现出和谐、宁静之美（图3-4-1，图3-4-2）。

　　Screen, a kind of furniture used to shield the wind inside traditional Chinese buildings, is called "screen the wind". As an important part of traditional furniture, the screen has a long history. It is generally displayed in a prominent position indoors, which plays a role in separation, beautification, windshield and coordination. Screen and classical furniture reflect each other, bring out the best in each other, and blend well, showing a harmonious and quiet beauty （Figures 3-4-1 and 3-4-2）.

图3-4-1　缂丝四堂花卉屏风
Kossu Screen With Four Shields of Flowers

图3-4-2　缂丝立屏百子图
Vertical Kossu Screen with "A Hundred children" Design

2. 缂丝屏条/Hanging Scrolls in Kossu Fabrics

屏条是中国书画装裱的一种式样，由于画身狭长，故能装裱成屏条形式。屏条单独挂的称"条屏"（屏条），四幅并排悬挂的称"堂屏"或"四季屏"。亦有四幅以上多至十二幅甚至十六幅，紧挂相连，成双数的完整画面，称"通屏"（图3-4-3）。

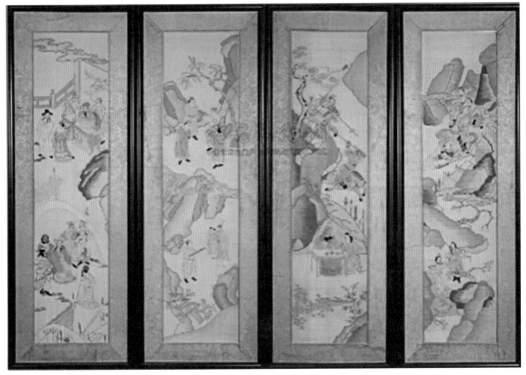

图3-4-3 缂丝三国人物屏条
Kossu Screen Strips of Three Kingdoms' Characters

Hanging scroll is a style of mounting Chinese paintings and calligraphy. Due to the painting's body being long and narrow, it can be mounted into hanging scroll form. Screen strips hung separately are called "strip screens" (hanging scrolls), and four screens hung side by side are called "hall screens" or "four seasons screens". There are also more than four pictures, as many as twelve or even sixteen pictures, which are closely linked and form even-numbered complete pictures, which are called "panoramic screen" (Figure 3-4-3).

3. 缂丝中堂/Central Scroll in Kossu Fabrics

缂丝中堂是挂在厅堂正中的大幅字画。一般是以中堂"正中央"显眼位置，悬挂四尺整张的"福""寿""龙""虎"等吉祥寓意的大字，再在左右配上"对联"，也有悬挂祖训、格言、名句书法题字或者祖先肖像、山水、老虎画的。

Central Scrolls are large calligraphy and painting hanging generally in the center of the hall in an outstanding position in the "center" of nave. Four feet of auspicious characters such as "blessing" "longevity" "dragon" "tiger" are hung, and then "couplets" are matched on the left and right. There are also having ancestral precepts, maxims, calligraphy inscriptions of famous sentences, or portraits of ancestors, landscapes and tiger paintings.

4. 缂丝台屏/Table-screen in Kossu Fabrics

缂丝台屏是古代书房重要的陈设品，置于书案之上，装饰客厅、书房等。台屏起装饰作品，需有装饰之用，但又不能过于耀眼，因此台屏以清雅为主。图3-4-4所示为牡丹题材的缂丝台屏，牡丹设色清雅，花朵绚丽绽放。

Table-screen in Kossu Fabrics were the important furnishings of the ancient study placed on the bookcase, decorating the living room, study room, etc. Table-screen needs to be used for decoration but they should not be too dazzling. Thus, it is mainly elegant. As shown in Figure 3-3-4, it is a kossu table-screen with peony which is elegant in color and flowers bloom brilliantly.

图3-4-4　缂丝台屏
Table-screen in Kossu Fabrics

二、缂丝椅披/Chair Cover in Kossu Fabrics

椅披为高级陈饰品，点缀于豪宅厅堂，缂丝椅披更是难得一见。明万历时期缂丝鼎盛，开创了许多新品种，缂丝椅披也是这个时期的创新品种。

Chair covers are high-grade ornaments, dotted in the halls of luxury houses, and kossu chair covers are even more uncommon. During the Wanli period of the Ming Dynasty, kossu fabric was popular and many new varieties were created, and the kossu chair cover was also an innovative variety in this period.

原耕织堂收藏有一件龙凤纹椅披，纵179cm，横63.5cm。应该是为宫廷或贵族专门定制的，搭背部分缂五彩福寿纹，椅背缂金夔龙椅，椅面海棠形开光内缂五彩祥凤一只，椅前垂饰缂织瑞兽、寿山福海纹（图3-4-5）。清代缂丝椅披也有织造，样式和装饰技法与明代相似，除了皇室、贵族外，高级官僚和富商大甲也享用这种奢侈的装饰用品（图3-4-6）。

There was a dragon and phoenix chair cover collected in the original Farming Hall, which was 179 cm in length and 63.5 cm in width. It should be specially made for the court or aristocrats, with colorful longevity patterns on the back, golden dragon chairs on the back of the chair and colorful auspicious phoenix on the face of the chair, weaving auspicious beasts and patterns which represent longevity and blessing on the front of the chair（Figure 3-4-5）. In the Qing Dynasty, chairs with kossu fabric were also woven, and their styles and decorative techniques were similar to those of the Ming Dynasty. Besides the royal family and nobles, senior bureaucrats and wealthy businessmen also enjoyed this luxurious decoration（Figure 3-4-6）.

图3-4-5 缂丝龙纹椅披
Kossu Chair Cover with Dragon Pattern

图3-4-6 明代缂丝龙纹椅垫
Kossu Chair Cushion with Dragon
Pattern in the Ming Dynasty

三、缂丝包首/Kossu Fabrics for Mounting

从唐代开始缂丝用于书画的装裱，按中国传统的装裱采用丝织品保护书画。宋代宫廷把唐、五代及以前的书画作品收集起来精心装裱，数量相当可观。缂丝是装裱用的首选材料，最珍贵的作品用缂丝，其次用锦绫等材料。辽宁省博物馆藏完整的《紫鸾鹊谱》，当时缂丝《紫鸾鹊谱》大量用于书画的装裱，但是不按照装裱尺寸织造，而是先织大幅后裁剪。

Kossu fabrics, has been used for mounting paintings and calligraphy since the Tang Dynasty, and silk fabrics are used to protect paintings and calligraphy according to Chinese traditional mounting. The court of Song Dynasty collected paintings and calligraphy works from the Tang, Five Dynasties and before, and carefully mounted them with a considerable number. Kossu fabric is the first choice for mounting, and the most precious works are used by kossu fabric, followed by brocade and silk. Liaoning Provincial Museum has a complete collection of *Phoenixes and Magpie with Purple Background* which was widely used for mounting paintings and calligraphy in the Song dynasty. It was first woven and then cut rather than was woven using the mounting size.

"鸳鸯戏莲"寓意夫妻和睦、相亲相爱之意，是明代常用的装饰题材，宋代绘画《唐人春宴图》的包首——缂丝《鸳鸯戏莲纹包首》为明代典型的装裱用缂丝（图3-4-7）。

图3-4-7　缂丝《鸳鸯戏莲纹包首》
Kossu *Mandarin Ducks Playing in the Lotus Pond*

"Mandarin ducks Playing in the Lotus pond" was the symbol of harmony and love between husband and wife, which was a common decorative theme in the Ming Dynasty. Head of package for kusso mounting of the Song Dynasty painting *spring, Banquet of Tang People-Mandarin Ducks Playing in the Lotus Pond*, was a typical mounting package in the Ming Dynasty（Figure 3-4-7）.

◎ 思考题/Questions for Discussion

1. 缂丝屏类包括哪些？/What are the screen types in kossu fabrics?
2. 简述缂丝包首的历史。/Introduction the history of kossu fabric for mounting.

◎ 第四章 / 缂丝的传承与创新
Inheritance and Innovation of Kossu Fabrics

◎ 概述/Introduction

通过前三章的学习，学习者对缂丝的历史渊源、技法工艺及种类有了清晰的了解，本章从缂丝的家族式传承、师徒传承、企业传承、学院传承等方面讲述缂丝的传承方式，进而引出缂丝的创新类型，学习者通过学习题材创新、呈现方式创新、材质创新、技艺创新、织机发展等知识，培养自身的创新意识，提高开拓进取的能力，并具备审美能力。

Through the study of the first three chapters, learners have a clear under-standing of kossu's historical origin, techniques and craft, and types. This chapter describes the inheritance method of kossu, including family inheritance, apprenticeship inheritance, enterprise inheritance, and college inheritance, and then leads to the innovative type of kossu. By learning the subject innovation, presentation method innovation, material innovation, technical innovation, loom development and other knowledge, learners can cultivate their own innovative consciousness, improve their ability to forge ahead, and have aesthetic ability.

◎ 思维导图 /Mind Map

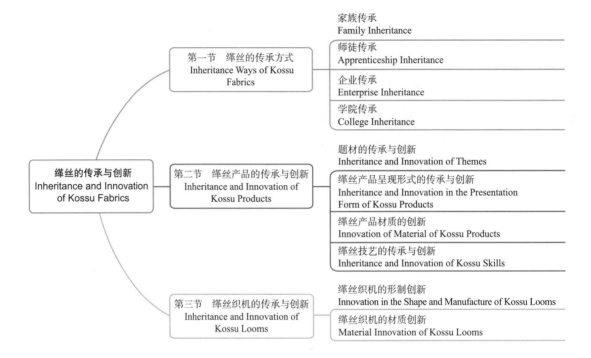

◎ 教学目标 /Teaching Objectives

知识目标 /Knowledge Goals

1. 掌握缂丝创新的类型 /Types of Innovation of Kossu Fabrics
2. 掌握缂丝传承的方式 /Inheritance Ways of Kossu Fabrics
3. 掌握缂丝产品创新的形式 /Innovation Forms of Kossu Products
4. 了解缂丝织机的发展 /The Development of Kossu Looms

技能目标 /Skill Goals

能根据缂丝产品的创新形式尝试创新 /Innovating Kossu Products According to Forms of Kossu Fabrics

素质目标 /Quality Goals

1. 具备良好的审美能力 /Possessing Good Aesthetic Ability
2. 能树立创新意识和创新精神 /Developing Innovative Consciousness and Spirit

思政目标 /Ideological and Political Goals

弘扬缂丝技艺和文化 /Promoting Kossu and Culture Skills

第一节 缂丝的传承方式/Inheritance Ways of Kossu Fabrics

缂丝织造技艺的传承方式有家族式传承、师徒传承、企业传承、学院传承四种方式。

缂丝传承

There are four inheritance ways of kossu skills: family inheritance, apprenticeship inheritance, enterprise inheritance, and college inheritance.

一、家族传承/Family Inheritance

家族传承是家庭的代代传承，属于活态传承，用动态的方式传承文化。通过家族代代人的努力，尽可能保持被传承文化的原貌及活力，以家族为传承主体，使家族传承具有血缘性、广泛性和永久性的特点。如王嘉良缂丝世家工作室的传承方式就以家族传承为主，已经历经六代，但是已经打破"传男不传女"的规矩，图4-1-1所示为缂丝世家网站，图4-1-2所示为缂丝世家第六代传人王建江手持祖父在天安门拍的照片。

Family inheritance refers to inheritance from generation to generation within the family, which belongs to living inheritance. Therefore, the culture of kossu skills is inherited dynamically. Through the efforts of generations of families, the original appearance and vitality of the inherited culture are maintained as far as possible. Family inheritance is characterized by a blood relationship, universality and permanence owing to a family is the main body of inheritance. For example, the transmission mode of Wang Jialiang family studio of kossu fabrics is mainly family inheritance, which has lasted for six generations. The rule of "passing on boys rather than girls" has been broken. Figure 4-1-1 shows the website of kossu family and Figure 4-1-2 shows Wang Jianjiang, the sixth generation of the inheritor, holding his grandfather's photo in Tian'anmen Square.

图4-1-1 缂丝世家网站
Website of Kossu Family

图4-1-2 第六代传人王建江手持祖父在天安门拍的照片
Wang Jianjiang, the of the Inheritor of the Sixth Generation,
Holding Grandfather's Photo in Tian'anmen Square

二、师徒传承/Apprenticeship Inheritance

师徒传承即个人通过传统师徒制的形式收取徒弟，耳提面命，带徒传艺。传统师徒的优势在于师徒沟通密切，对传习技艺来说是简单且有效的方式。如王金山工作室采用师徒传承的方式（图4-1-3）。然而，缂丝技艺本身学习周期较长，具有"易学难精"的特点；缂丝艺人带学徒费用高；从业人员老龄化现象严重；师徒传承所培养的人员有限。

Apprenticeship inheritance means that individuals recruit apprentices in the form of a traditional mentoring system and lead them to pass on kossu skills. The advantage of traditional mentoring lies in close communication between masters and apprentices, which is a simple and effective way to pass on skills. For example, the Studio of Wang Jinshan adopts the method of mentoring inheritance, as shown in Figure 4-1-3. However, the kossu skill itself takes a long time to learn, which is characterized by "be easy to learn but difficult to master." Moreover, the cost for artists to lead apprentices is so high. In addition, there are serious problems such as the aging of employees and limited personnel trained by mentoring succession.

图 4-1-3　王金山大师工作室
The Studio of Wang Jinshan

三、企业传承/Enterprise Inheritance

企业传承指的是企业培训员工应有的专业技艺，员工通过在企业里工作而进一步强化这种专业技艺，从而达到传承该技艺的目的。例如，"非物质文化遗产生产性保护基地"苏州市工业园区仁和织绣有限公司，人数在四五十人，不仅创新缂丝技艺，并在缂丝材料、产品研发、文化传承等方面作出一定的贡献。

Enterprise inheritance refers to how employees are trained professional kossu skills by the enterprise. Employees further strengthen their professional skills through working in the enterprise so as to achieve the purpose of inheriting the skills. For example, The Renhe Weaving and Embroidery Limited Company, a "productive protection base of intangible cultural heritage" in Suzhou Industrial Park, has 40 to 50 employees. The company not only innovates the kossu skill but also makes great contributions to materials, products, scientific research and cultural inheritance.

四、学院传承/College Inheritance

学院传承指的是通过学院里的正规教育体制进行传承，传承人与传承人之间是师生关系，分为"师生制""导师制""现代师徒制"等不同的学院传承模式。"现代师徒制"是一种通过学校、企业的深度合作与教师、师傅的联合传授，以培养学生技能为主的现代人才培养模式。例如，苏州技师学院与王金山大师工作室于2013年签订协议，苏州市非物质文化遗产办公室为提供专项资金，并对缂丝专业继承人就业、创业提供政策及资金扶持；王金山大师则提供人才培养的专任教师师资保障，并负责缂丝继承人考核选拔及提供实习、就业平台；苏州技师学院建立缂丝专业实践性人才培养基地，有针对性地培养缂丝工艺从业人员。尽管招生不多，但这种创新的人才培养模式打破了传统的非遗传承模式，是对缂丝传承的有益探索。苏州经贸职业技术学院以高等职业教育创新发展行动计划项目（2015—2018）缂丝技能大师工作室为平台，将缂丝课程融于人才培养方案，并组建缂丝社团和缂丝工作坊，实现了缂丝实践与理论教学比重，培养德技并修、知行合一的缂丝非遗传承人群。

College inheritance refers to the inheritance through the formal education system of the school. The relationship between inheritors is a teacher-student relationship, which can be divided into different school inheritance modes such as "teacher-student system""tutorial system" and "modern apprenticeship system.""Modern apprenticeship system" is a modern talent cultivation mode that focuses on cultivating students' skills through a deep cooperation between schools and enterprises and jointing teaching by teachers and masters. For example, Suzhou Technician College and the Studio of Wang Jinshan signed an agreement in 2013. Suzhou Intangible Cultural Heritage Office provided special funds, policy and financial support for employment and entrepreneurship of kossu craft and process. Master Wang Jinshan provided the full-time teachers for talent training, took charge of the examination and selection of weavers, and provided internship and employment platforms. Suzhou Technician College had set up a training base for practitioners of the kossu skills to train people engaged in the process. Although there were not many students enrolled, the innovative talent training mode broke the traditional non-inheritance mode which was a beneficial exploration for the inheritance of kossu skills. According to the Suzhou trade vocational and technical college in the development of higher vocational education innovation action plan (2015—2018), the studio of kossu skills as a platform, mixing courses of kossu skills into the talent training scheme, the college formed a community and workshops of kossu fabrics, which implemented the kossu proportion of practice and theory of teaching, training, skills and unity of kossu genetic population.

◎ **思考题**/**Questions for Discussion**

1. 缂丝传承的方式有哪些？/What are the inheritance ways of kossu fabrics?

2. 家族传承有什么特点？/What are the characteristics of family inheritance?

3. 什么是企业传承？/What is enterprise inheritance?

4. 师徒传承的方式是什么？/What is the way of inheritance from masters to apprentices?

5. 学院传承的方式是怎样的？/What is the way of college inheritance?

第二节　缂丝产品的传承与创新/Inheritance and Innovation of Kossu Products

一、题材的传承与创新/Inheritance and Innovation of Themes

（一）传统题材/Traditional Themes

缂丝产品的传承与创新

缂丝织造技艺能缂织各种装饰图案及风格多样的绘画作品，从传统的几何纹样、折枝花卉到书法、绘画作品，以及人物、风景、花草鱼虫、翎毛走兽都可以缂织，归纳起来主要集中在花鸟、走兽、风景、书法、人物、历史故事和宗教这七大类。

Kossu skills can weave art paintings and all sorts of adornment design. Diversified styles of painting include traditional geometric patterns, fold branch flowers, plants, calligraphy, paintings, worms, figures, landscapes, animals and feathers. Therefore, the style mainly can be classified into seven categories: flowers and birds, animals, landscapes, calligraphy, characters, historical stories and religion. The following parts will show them one by one.

花鸟：牡丹、山茶、梅、兰、竹、菊、玉兰、喜鹊等。花鸟题材历来是缂丝的主要装饰内容。如宋代缂丝作品有缂丝《梅鹊》《牡丹》《榴花双鸟》《凤凰牡丹》《山茶图》。

Flowers and birds: the subcategory includes peony, camellia, plum, orchid, bamboo, chrysanthemum, magnolia, magpie, etc. The theme of flowers and birds has always been the main decoration content of kossu fabrics. For example, works of kossu fabrics in the Song Dynasty included the *Plum Magpie, peony, durian flowers and two birds, phoenix and peony* and *camellia.*

走兽：龙、凤、仙鹤、瑞兽等。如宋代缂丝《紫鸾鹊谱》，以紫色熟丝作地，彩色纬丝织花卉鸾鹊。图案每组由五横排花鸟组成，文鸾、仙鹤、锦鸡等九种珍禽异鸟翩翩飞舞，和鸣翱翔。

Animals: there are dragons, phoenixes, cranes, auspicious beasts, etc. The kossu product of *Phoenixes and Magpie with purple Background* in the Song Dynasty used purple ripe silk as the ground, and colorful weft silk weaved flower and magpie. Each group is composed of five horizontal rows of flowers and birds, including nine rare and propitious birds such as phoenixes, cranes and golden pheasant.

风景：亭台楼阁、山水树木。清代书画鉴赏家卞永誉形容宋代风景缂丝："宋缂丝仙山楼阁，文绮装成、质素莹洁、设色秀丽、界画精工、烟云缥缈、绝似李思训（唐代著名画家）。"

Landscapes: there are pavilions, mountains and trees. Bian Yongyu（A connoisseur of painting and calligraphy in the Qing Dynasty）described the scenic silk as "it looks like Li Sixun (a famous painter in the Tang Dynasty) in a pavilion on a mountain. It is dressed in wen qi(a kind of beautiful silk product), which has bright and clean quality, beautiful colors and exquisite boundary paintings."

人物：三星、飞天侍女、八仙、仕女等。人物是丝织纹样中运用比较少的题材，然而却十分生动，有独到的艺术魅力。缂丝能缂织出人物的脸部线条、衣褶纹饰，所以缂丝中摹缂人物书画作品的不少。作品如缂丝《青牛老子图》《七夕图》《三星图》《八仙庆寿》。

Characters: there are three stars, a flying maid, eight immortals, maids and so on. The theme of characters is used less in the silk weave pattern, but they are very vivid and have unique artistic charm. The skill of kossu fabrics can weave the facial lines and pleats of characters, so many calligraphic and calligraphic works can be found in kossu products. Works are as follows: *Natural Cattle and Lao Tzu*, *Chinese Valentine's Day*, *Three Star* and *Eight Immortals to Celebrate Birthday*.

历史故事：如《后赤壁赋》《祝寿图》经典故事情节。

Historical stores: there are classic stories, such as *Post-Red Cliff Ode* and *Congratulation on the Birthday*.

宗教：早在唐代，就有用缂丝制品装点佛经，制作僧人的袈裟等，传世品中，元代就出现了缂丝佛画，一直流传至今。

Religion: as early as the Tang Dynasty, kossu products were used to decorate Buddhist sutras and make monk's cassocks, etc. Among the handed-down works, Buddhist paintings of kossu fabrics appeared in the Yuan Dynasty and have been coming down to the present.

（二）当代题材 /Contemporary Themes

缂丝工艺品从实用性向纯欣赏性的转变，也增加了缂丝工艺的表现题材，而且随着技艺的不断创新、完善，以及缂丝工艺与其他工艺的结合运用，缂丝题材已从最初的程式化的实用图案、佛像、佛经发展到山水、花鸟、人物、书法、建筑以及当代绘画、油画等。

Kossu crafts have changed from practicability to pure appreciation, which has increased the subjects of kossu fabrics. With the innovations and improvement of kossu skill, combined with other skills, kossu subjects have developed from the practical design of the stylized figure, buddha, the Buddhist scriptures into landscapes, flowers and birds, figures, calligraphy, architecture and contemporary painting, oil painting, etc.

例如，祯彩堂开发的苏州园林系列钱包，即将苏州园林式的风格建筑用缂丝技艺勾勒在钱包之上，图案简洁明了、用线果断，将苏州的建筑特色与缂丝技艺巧妙地融合在一起，更加凸显了独属于吴中的地域文化特色（图4-2-1）。再如，杭州西湖十景系列缂丝收纳用品，纹样选自当代画家陈家泠的西湖十景，图案包括苏堤春晓、断桥残雪、曲院风荷、花港观鱼、双峰插云、三潭印月、平湖秋月等。产品有名片夹、晚宴包、随身封、手提包、马蹄包、小盒包等（图4-2-2）。

For example, the Suzhou garden wallet series developed by the Zhencai Hall (a brand of kossu product) that used the kossu skill to outline the building of Suzhou garden on the wallet. The design was simple and clear with a decisive line, which skillfully integrated the architectural characteristics of Suzhou and the kossu skill to highlight the regional cultural characteristics unique to Wuzhong District, as shown in Figure 4-2-1. Besides, The ten Sceneries of West Lake in Hangzhou series of kossu storage products were selected from the ten sceneries of West Lake by Chen Jialing, who is a contemporary painter. The patterns included spring dawn on Su causeway, lingering snow on the broken bridge, wind lotus in winding courtyard, fish watching at flower harbor, two peaks through the cloud, three pools mirroring the moon, and autumn moon on a flat lake. Their products included cardholders, dinner bags, carry-on bags, handbags, horseshoe bags, small box bags, etc. (Figure 4-2-2).

图4-2-1 缂丝钱包——苏州建筑系列   图4-2-2 缂丝收纳——杭州西湖十景系列
Kossu Purse—Suzhou Architecture Series   Kossu Storage Products——Ten Views in West Lake

缂丝织造技艺经历了数千年的变化，题材越来越丰富，体现了当代缂丝艺人对作品品质的追求和创新，他们身上凝聚着缂丝人的精益求精、追求卓越的工匠精神，同时更是体现了只有创新才能使缂丝织造技艺重现辉煌！

The kossu skill has undergone thousands of years of changes with more and more diversified subjects, which reflects the pursuit and innovation of the contemporary kossu artists for the quality of their works. They embody the craftsmen's spirit of striving for perfection and pursuing excellence. And at the same time, it also shows that only innovation can make kossu skills return to glory.

二、缂丝产品呈现形式的传承与创新/Inheritance and Innovation in the Presentation Form of Kossu Products

缂丝产品呈现形式除了传统的缂丝服饰、丝巾、鞋履、团扇等日用品的形式外，还有缂丝挂屏、屏风、画册等观赏性缂丝，这些产品是传统缂丝的形式。如今，缂丝艺术在现代服饰、服饰品、乐器等各个领域进行了艺术创作和延展，令人耳目一新。

The ornamental and traditional form of kossu products includes hanging screens, screens, picture albums and some fabric forms of clothing, such as silk scarves, shoes, and fans for daily life. Today, kossu art has carried out artistic creation and extension in various fields such as modern clothing products and musical instruments, which is refreshing and fascinating.

（一）缂丝艺术在现代服装中的应用/Application of Kossu Art in Modern Clothing

目前，缂丝艺术在高级定制服装中的跨界合作已经取得一定成就，主要体现在设计师会充分发挥缂丝文化和缂丝的图案特色，将缂丝与服装恰如其分地融合在一起，形成具有现代审美趣味的华服。例如，缂丝世家第六代传人王建江与海派旗袍设计师苗海燕跨界合作推出《千里江山翠履霓裳》，如图4-2-3所示，该作品以宋代王希孟笔下的《千里江山图》为蓝本，礼服上半身采用缂丝制作如图4-2-3（a）所示，图案为云气山水，树木流水。款式上，上半身为肚兜，下半身为层层叠叠的流水造型，充分展示了东方美感。

At present, certain achievements have been made in the cross-border cooperation of kossu art in houte couture, which is mainly reflected in the fact that designers will give full play to the kossu culture and the pattern characteristics of kossu fabrics to appropriately integrate kossu skills with clothing to form magnificent clothes with modern aesthetics. For example, *the Garment of Vast Land* was launched by Wang Jianjiang the sixth-generation decendant of the kossu family and Shanghai-style Chinese dress designer Miao Haiyan which was cross-border cooperation, as shown in Figure 4-2-3. The product was based on the painting of *Vast Green Land* by Wang Ximeng in the Song Dynasty. The upper body of the dress was made of kusso fabric as shown in Figure 4-2-3(a). The patterns were gas landscape, trees, water. The Oriental aesthetic feels adequate that upper body was bellyband and the flowing water model of inferior half body layer was stacked.

 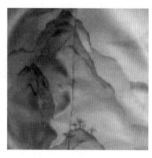 

(a)缂丝礼服上半身
Upper Part of the Kossu Dress

(b)细节图
Details of the Kossu Dress

(c)礼服全貌
Overall View of the Dress

图4-2-3　缂丝《千里江山翠履霓裳》
Kossu Dress of *Vast Green Land*

（二）缂丝艺术在鞋履中的应用/Application of Kossu Art in Shoes

缂丝艺术同样也应用在高级定制鞋履中。例如，缂丝世家第六代传人王建江与鞋履设计师李丹合作，推出缂丝高定女鞋（图4-2-4）。鞋面流畅的线条和丰富的色彩充分展示了缂丝表达图案色彩与线条的优势，同时这双女鞋将皮、丝、水晶等各种材质融合使用，营

图4-2-4　缂丝鞋履
Kossu Shoes

造远山树影的丰富层次、立体灵动效果。缂丝艺术在高定女鞋中的应用，充分展示了缂丝艺术良好的活态传承，同时体现了人们对优秀文化的热爱和推崇。

Kossu art is also used in haute couture shoes. For example Wang Jianjiang, the sixth-generation descendant of the kossu family, cooperated with shoe designer Li Dan to introduce kossu skills with high-set women's shoes, as shown in Figure 4-2-4. The smooth lines and rich colors outlined on the vamp fully demonstrate the advantages of kossu fabrics in expressing the colors and lines of the pattern.

Meanwhile, this pair of women's shoes used various materials such as leather, silk and crystal to create a rich hierarchical, three-dimensional and flexible effect of the shadow of distant mountains and trees. The application of kossu art in high-end women's shoes fully demonstrates that the kossu art has been well inherited, and at the same time reflects people's love and respect for excellent culture.

（三）缂丝艺术在箱包中的应用/Application of Kossu Art in Cases and Bags

缂丝箱包包括晚宴包、拎包、钱包、枕包、通勤包等。从题材上看，主要包括花卉、风景、建筑等。花卉题材包括写实花卉和装饰花卉。写实花卉如玉兰、睡莲等，比如祯彩堂与苏州博物馆合作，将沈周画作玉兰与缂丝结合，形成一系列女士箱包如图4-2-5（a）所示。装饰花卉如《梦回敦煌》系列晚宴包，将敦煌衣饰上的纹样和肌理通过缂丝的技艺展示出来［图4-2-5（b）］。风景题材如山水画等，如苏州工业园区仁和织绣有限公司将《忆江南·渡》景致应用于缂丝钱包中如图4-2-5（c）所示，十分秀美。缂丝箱包是缂丝服饰品中受众面较广的产品，在苏州博物馆文创商城的购买人数较多，足以看出人们对缂丝包的喜爱。

Kossu bags include dinner bags, handbags, wallets, pillow bags, commuting bags and so on. In terms of the subject, it mainly includes flowers, scenery and architecture. The theme of flowers includes realistic flowers and decorative flowers. Realistic flowers include magnolia, water lily, etc. For example, in cooperation with Suzhou Museum, Zhen Cai Hall had combined Magnolia, a painting by Shen Zhou, to form a series of women's bags, as shown in Figure 4-2-5 (a). For decorative flowers, it can be seen the dinner bag of *The Dream of Returning to Dunhuang* series shown in Figure 4-2-5 (b). The patterns and textures of Dunhuang clothing were displayed through the kossu skill. Scenic subjects include landscape painting and so on. For example, Renhe Weaving and Embroidery Limited Company of Suzhou Industrial Park applied the scene of *Remembering of The South* (a poet in the Tang Dynasty) into the kossu wallet, as shown in Figure 4-2-5 (c), which was exquisite. Kossu cases and bags are the most popular products that attract a wide range of people. There are a large number of people who buy kossu bags in the Cultural and creative Mall of Suzhou Museum, which shows people's love for kossu bags.

  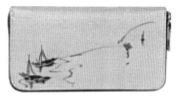

（a）沈周玉兰画在缂丝钱包中的应用
Application of Shen Zhou's Magnolia Painting in the Kossu Purse

（b）敦煌缂丝包
Kossu Handbag with Dunhuang Design

（c）《忆江南·渡》缂丝包
Kossu Hand bag with the Design of *Remembering of The South*

图 4-2-5　缂丝包
The Bag of Kossu Fabrics

（四）缂丝艺术在围巾、领带中的应用/Application of Kossu Art in Scarves and Ties

缂丝围巾多以素地为主，通过直经曲纬的特色展示缂丝围巾的纹理。例如，蔡霞明的缂丝围巾以汉代素纱禅衣为灵感，新创缂丝围巾薄纱清透飘逸，如图4-2-6（a）所示。又如改变缂丝围巾的材质，将羊绒与桑蚕丝结合制作缂丝围巾，面料柔软、手感丰满。缂丝领带主要是将独立花纹应用于领带中，配色比较低调、内敛。

Kossu scarves are mostly plain. The texture of kossu scarves is displayed by straight warp and curved weft. For example, Cai Xiamin's kossu scarf was inspired by the Pure Gauze Unlined clothes in the Han Dynasty. Cai Xiamin created a new kossu scarf with clear and elegant gauze , as shown in Figure 4-2-6 (a). Another example is to change the material of kossu scarf and combine cashmere and mulberry silk to make a scarf to feel soft and satisfied.The main form of kossu tie is to apply the independent pattern, and the color matching is low-key and restraining.

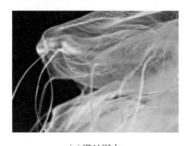 

(a)缂丝围巾　　　　　　　　　　　(b)缂丝领带
Kossu Scarf　　　　　　　　　　　Kossu Tie

图4-2-6　缂丝围巾及领带
Kossu Scarf and Tie

（五）缂丝艺术在首饰品中的应用/Application of Kossu Art in Jewelry

近年来，设计师将缂丝面料应用于首饰设计中，是缂丝创新产品之一。将缂丝工艺与首饰工艺结合，展示缂丝的图案，受到年轻人的喜爱，如图4-2-7所示。缂丝首饰包括戒指、耳坠、项链、胸针等，既有抽象的山水也有具象的花鸟，可谓小巧精致。

图4-2-7　缂丝首饰
Kossu Jewelries

In recent years, designers have applied kossu skills to the design of jewelry, which is one of the popular innovations of products. The process of design is combined with the jewelry skills to display the kossu pattern, which is highly sought after by young people, as shown in Figure 4-2-7. Kossu jewelry consists of rings, eardrops, necklaces, brooches, etc. With both abstract landscapes and figurative flowers and birds, it can be described as small and exquisite.

（六）缂丝艺术在茶具中的应用/Application of Kossu Art in Tea Set

茶具类缂丝产品的创意构想源于中国人自古对茶道的热爱，这是一种深入骨髓的民族情结与怀旧意识。祯彩堂负责人陈文看到了茶文化对于中国人生活习惯的影响，直接将缂丝技艺伸入到茶道文化之中，开发设计了缂丝茶席、便携式缂丝茶杯袋等产品。茶具类缂丝产品不仅仅是在设计与题材上的创新，更为重要的是拓展了缂丝在现代人日常生活领域的应用。外层为缂丝织造，无论从整体设计、用料材质、包装设计、价格定位等方面都展现了其高端精品的产品定位（图4-2-8）。

The creative idea of kossu products for tea sets originated from Chinese people's love for tea ceremony since ancient times, which is a kind of national complex and nostalgia consciousness deeply rooted in the marrow. Inspired by the influence of tea culture on the living habits of Chinese people, Chen Wen, the director of Zhencai Hall, directly introduced the kossu skill into the tea ceremony culture, developing and designing kossu tea mat, portable tea bag and other products. Kossu products of tea sets are not only innovations in design and subject matter, but more importantly, they expand the kossu fabric in the daily life of modern people. The kossu fabrics are the outer layer of the product, showing its high-end and high-quality product positioning in terms of the overall design, material, packaging design, price positioning and other aspects（Figure 4-2-8）.

图4-2-8 缂丝茶席
Kossu Products of Tea Set

（七）缂丝艺术在乐器中的应用/Application of Kossu Art in Musical Instruments

缂丝织物与吉他相结合是缂丝产品的又一创新，这一融合创新充分展示了缂丝应用的多样性和适应性，它可以与皮具结合，也可以与木材结合，使产品更具推广性（图4-2-9）。

The combination of kossu products and guitar is another innovation of kossu skills. The innovative integration fully demonstrates the diversity and adaptability of kossu skills. It can be combined with leather goods or wood, making the products more widely promoted（Figure 4-2-9）.

三、缂丝产品材质的创新/Innovation of Material of Kossu Products

随着纺织科技的发展，新材料的出现和加入给缂丝产品带来更多的功能性和特色。例如，2017年范炜焱带着5个系列12件缂丝家具作品获得2017意大利米兰国际家具展——青年明日之星沙龙展评审会特别奖。其中"记忆灯"可以随意塑形，能变成千纸鹤，也能变成一朵盛开的鲜花，它通过丝绸和金属记忆丝结合，再加上缂丝技艺制作而成（图4-2-10）。

图4-2-9　缂丝应用于吉他
Application of Kossu Art in Guitar

With the development of textile technology, the emergence and addition of new materials can bring more functionality and features to kossu products. For example, in 2017, Fan Weiyan won the special award of the jury of the 2017 Milan International Furniture Fair — "Young Star Salon Exhibition" with 5 series and 12 pieces of kossu furniture works. Among them, the work of *Memory Lamp* could be shaped at will and could be changed into a thousand paper cranes or a blooming flower. It was made by combining metal silk with kossu techniques, as shown in Figure 4-2-10.

(a)缂丝台灯　　　　　　　　　　　(b)缂丝吊灯
Kossu Lampstand　　　　　　　　　Kossu Ceiling Lamp
图4-2-10　缂丝灯
Kossu Lights

缂丝台屏《水墨青山》用强度极高的单丝，改变传统缂丝的原料和纱线，采用极细的丝线进行织造，形成新的视觉效果（图4-2-11）。

Ink Castle and Peak, the table screen of kossu fabrics, used extremely high strength monofilaments to change the raw materials and yarns of traditional kossu material and adopted extremely fine silk threads for weaving, forming a new visual effect, as shown in Figure 4–2–11.

此外，利用功能性后整理对缂丝面料进行处理，也是缂丝产品的创新手段之一。例如，根据缂丝织物防水、防油以及防污的要求，利用真丝低温三防材料，对缂丝三防测试技术的确定，形成的缂丝产品具备防水、拒油和易去污等功能。

图4-2-11 《水墨青山》缂丝台屏
Kossu Table Screen *Ink Castle and Peak*

In addition, the treatment of kossu fabrics by functional finishing is also one of the innovative means of kossu products. For example, according to the requirements of water–proof, oil–proof and pollution–proof kossu fabrics, the kossu product has the functions of waterproof, oil–proof and easy decontamination through the determination of the test technology of the three–proof material with low temperature and pure silk.

缂丝与其他材质的配合，也是缂丝产品材质创新的途径之一。例如，缂丝包，将缂丝与牛皮结合，创造出具有现代风格特征的缂丝箱包。

The combination of kossu fabrics and other materials is also one of the ways of material innovation of kossu products. Taking kossu bags, for example, kossu skills are combined with cowhide to create bags with modern style features.

四、缂丝技艺的传承与创新/Inheritance and Innovation of Kossu Skills

目前，缂丝织造技艺在传承方面不仅达到缂丝鼎盛时期的水平，王金山、王嘉良等大师曾多次复制宋代、清代等各朝代的缂丝作品，并达到甚至超过当年的缂丝技艺水平。

At present, the kossu skill has not only reached the level of its heyday but also reaches or even surpasses the kossu skill of the day, such as masters of Wang Jinshan and Wang Jialiang, who have copied kossu works of the Song Dynasty, Qing Dynasty and other dynasties for many times.

如今，在技术发展的背景下，缂丝技艺也在创新。例如，万事利集团珍藏的《三潭印月》，运用了一种实验性的创新技法来改变缂丝原本紧密的组织结构，创新出了一种全新的水波纹肌理，中密外疏，半实半空，形成鲜明对比，完美地再现了三潭印月"湖中有深潭，明月印水渊，石塔来相照，一十八月圆"的奇异景致，如图4-2-12所示。

Today, with the development of technology, kossu skills are also innovating. With the collection *Three Pools Mirroring the Moon* of the Wensli Group, the innovation of the use of an experimental technique changed kossu fabrics originally tight organization structure, innovation out of the watering texture, which was a new kind of water ripple and thin outside half solid half empty. It perfectly reproduced the scenery that "the lake has deep pools, the bright moon shines on the lake, the stone tower stands face to face, the moon is round on the 18th day of the lunar calendar", as shown in Figure 4-2-12.

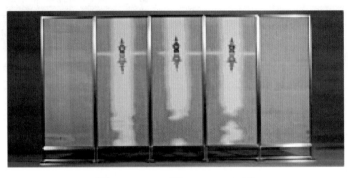

图4-2-12　缂丝《三潭印月》
The Kossu Product of *Three Pools Mirroring the Moon*

◎ 思考题/Questions for Discussion

1. 缂丝产品创新的种类有哪些？/What are the types of kossu product innovation?

2. 缂丝产品呈现形式的创新有哪些？/What are the innovation in the presentation form of kossu products?

第三节　缂丝织机的传承与创新/Inheritance and Innovation of Kossu Looms

一、缂丝织机的形制创新/Innovation in the Shape and Manufacture of Kossu Looms

缂丝织机

为更好地满足大产品的创作，通常要对缂丝织机门幅进行改造。例如，独幅缂丝作品《观无量寿经图》，宽4.6m、长4.8m，采用1600多种颜色丝线织就，从画图设计、造机到缂织完工，前后耗时10年，其中造机就耗时5个月，由于缂丝织物较宽，缂丝机的机身宽达到5.5m，如图4-3-1所示。

在造机时，需增加缂丝后轴的宽度，同时保证缂丝经线张力均匀。

To satisfy the creation of large products, the width of kossu loom is usually remolded. For example, *View of Infinite Longevity* was 4.6 m in width and 4.8 m in length and was woven with silk threads of more than 1600 colors. It took 10 years from drawing design, machine building to completion of kossu fabrics, of which the machine–building took 5 months. As the kossu fabrics were relatively wide, the body of kossu loom was 5.5 m wide, as shown in Figure 4–3–1. During manufacturing, the width of the rear axis should be increased and the tension of the warp should be uniform.

图4-3-1　缂丝大织机
Large Kossu Loom

同时，缂丝小织机也是缂丝织机的创新手段之一，为更好地在中小学中推广缂丝，陈文等人将缂丝织机改为折叠式缂丝织机，一是便于携带，二是满足小作品的实践与创作，图4-3-2所示为陈文老师带着小学生们利用便携式可折叠缂丝织机进行缂丝技艺实践。缂丝小织机的特征在于：包括铰接的前轴架和后轴架，具有收折和打开两个状态；前轴架上设有前轴杆及前张紧轴，对应前张紧轴设有前刹结构；后轴架上设有后轴杆，前轴架的下端设有后张紧轴，对应后张紧轴设有后刹结构；前轴架上铰接有立架，立架上端具有上转动杆；后轴架下端具有下转动杆。还包括两番片，悬设于上转动杆下方；第一番片对应排在单数位置的经线设置，第二番片对应排在偶数位置的经线设置；两番片的上横杆两端均通过上牵引绳与上转动杆两端连接，下横杆的两端通过下牵引绳与下转动杆的两端连接。还包括踏脚件，两端分别通过拉绳与下转动杆的两端连接。还包括筘箱架，上端固设有筘箱，底部转动连接于后轴架下端。

At the same time, the small kossu loom is also one of the innovative means of kossu loom.

For better publicity in primary and secondary schools, collapsible kossu loom was invented and promoted by Chen wen and some other people. It is not only easy to carry but also meets the demands of small work practice and creation. In Figure 4–3–2, there was a teacher with the pupils to use portable folding kossu loom for skill practice. The small kusso loom is characterized by a hinged front axle frame and rear axle frame, with folding and opening two states. The front axle frame is provided with a front axle rod and a front tensioning shaft, and the front tensioning shaft is provided with a front brake structure; the rear axle frame is provided with a rear axle rod, the lower end of the front axle frame is provided with a rear tensioning shaft, and corresponding rear tensioning shaft is provided with a rear brake structure. The front axle is hinged with a vertical frame, and the upper end of the vertical frame is provided with a rotating rod. The lower end of the rear axle frame is provided with a lower rotating rod. It also includes two plates suspended below the upper rotating rod. The first slice corresponds to the warp settings in odd positions, and the second slice corresponds to the warp settings in even positions. Both ends of the upper bar are connected with both ends of the upper rotating rod through the upper traction rope, and both ends of the lower bar are connected with both ends of the lower rotating rod through the lower traction rope. The two ends are respectively connected with both ends of the lower rotating rod by a rope. The sley box frame is also provided with a reed box fixedly at the upper end, and the bottom is rotationally connected to the lower end of the rear axle frame.

图4-3-2 用于小学生教学的缂丝小织机
Application of Weaving Loom in Primary School Teaching

二、缂丝织机的材质创新/Material Innovation of Kossu Looms

长时间使用的木制织机容易老化，因此有些缂丝厂对织机进行改造。因此，织机全部采用铁制，其结构和原理与木制织机基本相同。主要部分部件的改造，如将原有的竹筘改用钢筘，如图4-3-3所示。

Wooden weaving looms are easy to be antiquated and broken for a long time to use, therefore some kossu weaving factories turn to transform and improve the weaving looms. Therefore all looms are made of iron, and their structure and principle are the same as those made of wood. The main parts of the reconstruction include the original bamboo reed, using reed, as shown in Figure 4-3-3.

织机前轴的转动也采用齿轮转动来代替人工卷绕，一是使张力更加均匀，二是更省力。同时铁织机一般采用下番头，这样可使工艺简化（图4-3-4）。

The rotation of the front axle of the weaving loom also adopts gear rotation to replace manual winding. First, the tension is more well-distributed; moreover it saves effort. At the same time, iron looms generally use the lower head, thus simplify the process.

图4-3-3　钢筘
Steel Reed

图4-3-4　前轴
Front Axle

◎ 思考题/Questions for Discussion

1. 缂丝织机创新方式有哪些？/What are the innovative ways of kossu weaving loom?

2. 缂丝钢筘和竹筘的作用有区别吗？/Is there any difference between the functions of steel reed and bamboo reed?

◎ 第五章 / 缂丝非物质文化遗产传承人
Inheritors of Kossu Intangible Cultural Heritage

◎ 概述 /Introduction

在"保护、提高、发展"的方针指导下，缂丝艺术在各个方面取得了重大成就。传承古代缂丝艺术，在此基础上不断创新品种和技术，由老一代缂丝艺人悉心指导，培养了一大批新一代缂丝艺人，为缂丝艺术的持续发展储备了优秀的人才，甚至成为非物质文化遗产项目代表性传承人。非物质文化遗产项目代表性传承人，是指经国务院文化行政部门认定的，承担国家级非物质文化遗产名录项目传承保护责任，具有公认的代表性、权威性与影响力的传承人。

Under the guidance of the policy of "protection, improvement and development", the art of kossu fabric has made great achievements in various aspects. On the basis of inheriting the ancient kossu art, the varieties and technologies are constantly innovated. Under the careful guidance of the older generation of kossu artists, a large number of new generation kossu fabric artists have been trained, which has reserved outstanding talents for the sustainable development of kossu art and even become the representative inheritors of intangible cultural heritage projects. Representative inheritors of intangible cultural heritage projects refer to inheritors who are acknowledged by the cultural administrative department of the State Council, undertake the responsibility of inheriting and protecting national intangible cultural heritage list projects, and have recognized as representative, authoritative and influential.

本章根据任务要求，通过介绍非物质文化遗产传承人的基本信息、创新技法、代表作等，带领学习者走近缂丝艺人，体会并学习艺人身上的工匠精神，提升学习者的创新动力。

According to the task requirements, this chapter introduces the basic information, innovative techniques and representative works of kossu inheritors of intangible cultural heritage, leading learners close to the kossu fabric artists, experiencing and learning the craftsmanship spirit of artists, as well as enhances learners' innovative power.

◎ 思维导图/Mind Map

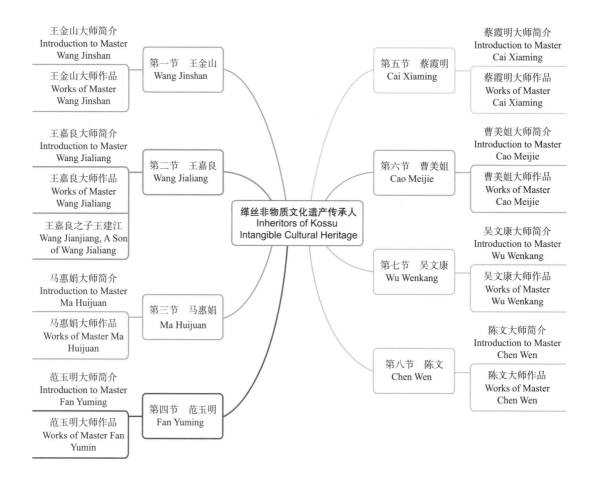

◎ 教学目标/Teaching Objectives

知识目标/Knowledge Goals

1. 了解缂丝非物质文化遗产传承人/Inheritors of Kossu Intangible Cultural Heritage

2. 掌握缂丝非物质文化遗产传承类别/Inheritance Categories of Kossu Intangible Cultural Heritage

3. 熟知近代缂丝创新技法及产品特点/Innovative Techniques and Product Characteristics of Modern Kossu Fabrics

4. 了解缂丝非物质文化遗产传承人代表作/Representative Works of the Inheritors of Kossu Intangible Cultural Heritage

技能目标/Skill Goals

1. 能掌握近代缂丝技法/Mastering Modern Kossu Techniques

2．熟知作品名称及含义 /Knowing the Works and the Connotations

素质目标/Quality Goals

1．具备传承弘扬非物质文化遗产的能力 /Acquiring the Ability to Inherit and Promote Intangible Cultural Heritage

2．具备吃苦耐劳的工匠精神 / Possessing Hard-working Craftsman Spirit

思政目标/Ideological and Political Goals

1．提升眼界和格局，扩大缂丝非遗文化的受众面 / Improving the Vision and Pattern, and Expanding the Audience of Kossu Intangible Culture Heritage

2．加强文化修养，树立文化自信 /Enhancing Cultural Cultivation and Establishing Cultural Self-confidence

第一节　王金山/Wang Jinshan

一、王金山大师简介/Introduction to Master Wang Jinshan

王金山大师

1．成长历程/Personal Development

王金山，1939年出生于苏州，1956年进入苏州刺绣工艺美术生产合作社（现苏州刺绣研究所），师从缂丝名人沈金水学艺，并先后向顾仲华、张辛稼、吴木、徐绍青、张继馨等画家学习书法绘画，历任苏州刺绣研究所缂丝技艺指导、苏州缂丝厂厂长、苏州工艺美术研究所副所长和苏州工艺美术博物馆（筹）馆长。

Wang Jinshan was born in Suzhou in 1939. In 1956, he joined Suzhou Embroidery Arts and Crafts Production Cooperative Association (later Suzhou Embroidery Research Institute). He learned from Shen Jinshui, a famous kossu artist, and studied calligraphy and painting from Gu Zhonghua, Zhang Xinjia, Wu Mu, Xu Shaoqing, Zhang Jixin and other painters successively. He has successively served as the kossu skill director of Suzhou Embroidery Research Institute, the director of Suzhou Kossu Factory, the deputy director of Suzhou Arts and Crafts Research Institute and the curator of Suzhou Arts and Crafts Museum (to be established).

2．对缂丝的研究/Research on Kossu Fabrics

王金山大师始终以发展缂丝艺术为己任。他悉心培育缂丝后继人才，在苏州刺绣研究所、苏州缂丝厂工作期间先后培养了多名学生，被认定为非物质文化遗产国家级代表性传承人后更是致力于培养年青一代，患病之后仍孜孜不倦地传授技艺。他曾两次亲任苏州缂丝研究会会长，为缂丝事业的提升发展耗尽了心血。他对宋、元、明、清历代缂丝艺术有深入的研究并融会贯通，尤其是戗色表现技巧能顺应画理要求，随画敷梭。他创新的无规则随形任意变化晕色法，丰富了色彩变化层次，达到笔精墨妙、神韵俱足的艺术效果。

Master Wang Jinshan always takes the development of kossu art as his own duty. He carefully cultivated the successors of kossu, and trained many students during his work in Suzhou Embroidery Research Institute and Suzhou Kossu Factory. Recognized as a national representative inheritor, he devotes himself to training the younger generation, and continues to teach skills tirelessly after his illness. He once personally served as the president of Suzhou Kossu Research Association twice, and made painstaking efforts for the promotion and development of kossu fabrics. He has a deep research on the art of kossu fabrics in the Song, Yuan, Ming and Qing Dynasties, especially the performance skills of draw weaving which can conform to the painting requirements and apply shuttle with painting. His innovative demitint method of randomly changing colors and shapes enriches the layers of color, and achieves the artistic effect of exquisite brushwork and full charm.

他对缂丝史论有深入研究，先后撰写《缂丝技艺发展》《论宋缂丝技艺表现手法》等论文和《苏州缂丝》专著。《缂丝生产工艺流程及缂丝特点》《缂丝技术及制作工艺》等论文刊入《中国科学技术史》《中国传统工艺全集》。

He has made an in-depth study on the history of kossu fabrics, and has successively written academic papers such as *Development of Kossu Skills*, *On the Performance and Techniques of Song Kossu Skills* and a monograph *Suzhou Kossu Fabrics*. Papers such as *Kossu Fabrics Production Process and Characteristics* and *Kossu Fabrics Technology and Production Technology* have been selected and published in *The History of Chinese Science and Technology* and *The Complete Collection of Traditional Chinese Crafts*.

3. 成就与荣誉/Achievements and Honors

半个多世纪以来，王金山继承优秀的传统技艺，并经努力探索和创新，取得了突出的成绩。他被评为国家级非物质文化遗产代表性传承人，多次被评为部、省劳动模范以及先进生产（工作）者，并被国务院授予国家级突出贡献专家，被世界手工艺联合会授予"亚太地区手工艺大师"称号。央视"东方时空""东方之子""传承·大师"特别系列节目和央视"探索发现""寸丝寸金"等栏目详细地介绍了他的突出业绩。他为人正直、胸怀坦荡、生活简朴、淡泊名利，具有艺术家的人格风范，深受工艺美术界广大同仁的敬重与爱戴。

For more than half a century, Wang Jinshan has inherited excellent traditional skills, made great efforts to explore and innovate, and made outstanding achievements. He has been rated as a national representative inheritor of intangible culture, a ministry and provincial labor model and an advanced producer (worker) for many times, and he has been awarded a national expert with outstanding contributions by the State Council, as well as the title of "Asia-Pacific Handicraft Master" by the World Handicraft Federation. His outstanding achievements have been introduced in detail by CCTV's special series programs "Eastern Horizon" "Oriental Horizon" and "Inheritance and Master", as well as CCTV's programs "Exploration and Discovery" "Inch Silk and Inch Gold" and other columns. He is upright, open-minded, simple in life, indifferent to fame and fortune, and has the personality of an artist, which made him deeply respected and loved by the vast number of colleagues in the arts and crafts industry.

二、王金山大师作品/Works of Master Wang Jinshan

1960年缂织的宋徽宗《柳鸭芦雁图》获苏州市工艺美术优秀创作二等奖。1963年派往北京故宫博物院复制南宋缂丝名家沈子蕃的《梅花寒鹊图》等缂丝作品，被故宫博物院列为国家一级文物收藏。1970年缂织的《金地牡丹屏风》《花篮》作为国礼赠送美国和日本，领衔创制的巨幅金地书法缂丝诗词《西江月·井冈山》创新了金线与玄色线相结合的绞花线新技法，收藏于毛主席纪念堂西大厅。1980年创制的缂丝三异（异色、异样、异织）《牡丹·山茶·双蝶》，全异缂丝《寿星图》，作为珍品收藏于中国工艺美术馆。1990年，复制的宋缂丝《紫芝仙寿图》由苏州工艺美术博物馆陈列展出。21世纪以来，为首都博物馆修补破损400多处的清乾隆缂金十二章龙袍，缂织的《金地牡丹蝴蝶中堂》参加文化部举办的中国非物质文化遗产保护成果展，并被中国艺术研究院收藏。王金山复制的著名的作品包括朱克柔的《莲塘乳鸭图》、沈子蕃的《青碧山水图》《梅花寒鹊图》等，如图5-1-1~图5-1-3所示。

In 1960, *Willow Duck and Wild Goose* of Song Emperor Huizong won the second prize of Suzhou Arts and Crafts Excellent Creation. In 1963, he was sent to the Palace Museum in Beijing to reproduce the kossu works such as *Plum Blossoms and Magpies* by Shen Zifan, a famous kossu artist in the Southern Song Dynasty, which was listed as a national first-class cultural relic collection by the Palace Museum. In 1970, *Golden Field Peony Screen* and *Flower Baskets* were presented as national gifts to the United States and Japan. The poem *West Lake Moon and Jinggang Moutain*, a giant gold kossu calligraphy created by the team leader Wang Jinshan, has innovated the new technique of twisting flower threads that combines gold threads with dark black threads, and is collected in the West Hall of Chairman Mao Memorial Hall. In 1980, *Peony, Camellia, Double Butterflies* created with three different kossu techniques (different colors, different patterns and different weaving) and *Longevity,* a kossu work with completely different designs in both sides, were listed as treasures and collected in China National Arts & Crafts Museum. In 1990, his reproduced Song Kossu work *Purple Ganoderma, Immortals and Longevity* which was exhibited by Suzhou Arts and Crafts Museum. Since the beginning of 21 century, he has repaired more than 400 pieces of damaged golden dragon robes of Qianlong Period of the Qing Dynasty for Capital Museum, and his woven *Jindi Peony Butterfly Nave* has participated in the Exhibition of Chinese Intangible Cultural Heritage Protection Achievements organized by the Ministry of Culture, and has been collected by China Academy of Art. All his life, Wang Jinshan reproduced lots of famous works in history, including Zhu Kerou's Suckling *Ducks in the Lotus Pond*, Shen Zifan's *Green Landscape* and *Plum Blossoms and Magpie* in the Song Dynasty, as shown in Figures 5-1-1 to 5-1-3.

图 5-1-1　缂丝《莲塘乳鸭图》
Kossu *Suckling Ducks in the Lotus Pond*

图 5-1-2　缂丝《梅花寒鹊图》
Kossu *Plum Blossoms and Magpies in Wonter*

图 5-1-3　缂丝《青碧山水图》
Kossu *Green Landscape*

◎ 思考题/Questions for Discussion

1. 1980年，王金山大师创制了缂丝三异，指哪三异？/Master Wang Jinshan created three differences in kossu weaving techniques in 1980. What are the three differences?

2. 王金山大师复制了朱克柔的哪幅作品？/Which work of Zhu Kerou was copied by Master Wang Jinshan?

3. 王金山大师复制了沈子蕃的哪些作品？ /What works of Shen Zifan were copied by Master Wang Jinshan?

4. 王金山大师的作品中被故宫博物院收藏的有哪些？/What are the works of Master Wang Jinshan collected by the Palace Museum?

第二节　王嘉良/Wang Jialiang

一、王嘉良大师简介/Introduction to Master Wang Jialiang

王嘉良大师

1. 成长历程/Personal Development

王嘉良，1939年10月出生于陆墓缂丝村，其祖上早在光绪年间就从事缂丝，并为皇室制作过袍服，其曾祖父王新事是同治和光绪年间的缂丝名匠，一度为宫廷匠师，光绪年间曾为慈禧太后寿庆缂织八仙庆寿袍料和霞帔。其祖父王锦亭，是晚清至民国初年的缂丝名匠，擅长丹青，早年曾制作清廷御用缂丝品，代表作缂丝《麻姑献寿图》参加1915年巴拿马国际博览会并获奖。其父王茂仙受聘于苏州刺绣工艺美术生产合作社任缂丝技术指导，为保护传承缂丝技艺、培养新一代接班人做出了贡献。

Wang Jialiang was born in October 1939 in the kossu village of Lu Tomb. His ancestors worked in kossu fabrics as early as in Guangxu period and made robes for the royal family. His great-grandfather Wang Xinshi was a famous kossu craftsman in the Tongzhi and Guangxu years, and was once a court craftsman. During the Guangxu period, he weaved the robes and capes of eight immortals to celebrate birthday of Empress Dowager Cixi. And his grandfather Wang Jinting, a famous kossu craftsman from the late Qing Dynasty to the early Republic of China, was good at painting. In his early years, he made kossu products for royal families. His masterpiece *Ma Gu's Celebration of Birthday* won an award in Panama International Expo in 1915. His father Wang Maoxian was employed by Suzhou Embroidery Arts and Crafts Production Cooperative Association as the technical director of kossu fabrics, and made contributions to rescuing and preserving the ancient art of kossu fabrics as well as training a new generation of successors.

王嘉良从小耳濡目染受家庭艺术熏陶，9岁从父学艺。1954年2月，王嘉良随父应招进苏州刺绣工艺美术生产合作社缂丝小组，成为该社年龄最小的职工，与更多前辈和同期学工交流，进步飞速，缂艺日臻。1974年后，王嘉良的缂丝技艺得到了社会的重视。浙江德清、江苏东台等地的工艺品生产企业，先后邀请其做缂丝技术指导，王嘉良毫无保留地传授技艺，把吴中古艺带到了浙江和苏北地区。1979年，进吴县缂丝总厂，先后任技术指导车间主任和缂丝研究室副主任，自此，王嘉良成为缂丝精品织造的骨干。

Wang Jialiang was influenced by the family art since childhood and learned art from his father at the age of 9. In February, 1954, Wang Jialiang joined the Kossu Group of Suzhou Embroidery Arts and Crafts Production Cooperative Association with his father, and became the youngest employee of this organization. He communicated with more seniors and fellow students and made rapid progress in his skills. After 1974, Wang Jialiang's kossu skills gained the attention of the society. Handicraft manufacturers in Deqing of Zhejiang Province, Dongtai of Jiangsu

Province and other places had invited him to give technical guidance on kossu fabrics, and Wang Jialiang had taught skills without reservation, bringing Wuzhong ancient art to Zhejiang and northern Jiangsu. In 1979, he entered Wu County Kossu General Factory and served as the director of technical guidance workshop and the deputy director of Kossu Fabrics Research Office. Since then, Wang Jialiang has become the backbone of fine kossu weaving.

2.对缂丝的研究/Research on Kossu Fabrics

王嘉良得祖传技法，并在陈阿多、徐祥山等前辈艺人的影响下刻苦钻研技艺戗法，精益求精，推陈出新，所缂品种繁多，均属精品。为了缂织人物的表情和眼神，他多次到号称"天下罗汉二堂半"之一的紫金庵，潜心琢磨罗汉的眼睛。他缂织的《云龙图》以泼墨画稿为粉本，采用戗法10多种、浓淡墨色17种，画面中的乌龙在惊涛拍浪之中，昂首云天，蓄势凌空，扶摇翻滚，气势磅礴，该作品在德国国际博览会展出获得好评；缂丝《金丝猴图》是一幅以写意国画为粉本的佳作，画中的金丝猴盘坐在虬松上，目视前方飞舞的小蜜蜂，猴尾下垂而末端呈上翘动势，形态自然逼真，惹人喜爱，富有中国水墨画的韵味；1986年，受江苏省丝绸进出口公司委托，成功研制缂丝"拉挂"立体画；1987年被评为工艺技师。

Wang Jialiang acquired ancestral techniques, and under the influence of predecessors such as Chen Aduo and Xu Xiangshan, he assiduously studied the techniques, kept improving and brought forth new ideas. Hence, various types and varieties of kossu fabrics have been made and all of them are top-grade products. In order to produce the characters' expressions and eyes, he went to Zijin Temple, one of the so-called "two and a half halls of arhats in the world", and devoted himself to pondering arhats' eyes. His *Yunlong Map* was based on ink-splashing paintings, using more than 10 kinds of methods and 17 kinds of shades of ink. In the picture, the black dragon was in the stormy waves, holding its head high in the sky, rolling with a tremendous momentum. His works have been well received in the German International Expo. *Golden Monkey picture* was a masterpiece based on freehand Chinese painting. In the painting, the golden monkey sat on the pine and looked at the bees flying in front. The monkey tail drooped and the end was warped, which was natural, realistic and attractive, and full of the charm of Chinese ink painting. In 1986, commissioned by Jiangsu Silk Import and Export Corporation, he successfully developed the three-dimensional picture of "pulling and hanging" of kossu products, and was rated as a craft technician in the following year.

退休之后的王嘉良，更加热爱缂丝，继续他的经纬生涯并在技艺上不断探索、不断创新。他凭借祖传技艺以及自己几十年的缂丝实践经验，逐渐形成了"王氏"缂丝艺术风格和特色：一是擅长缂织人物题材，其中人物开相自成一格，为业内人士称道；二是创造了三异缂丝新品；三是擅长龙袍绳织，龙袍可谓"王氏缂丝"的传统产品，王嘉良生活的年代，龙袍等御用缂丝品早已停产，他对龙袍没有感性认识。然而，当故宫博物院的专家慕名找到他时，他果断地承接了龙袍的复制，他凭借多年的缂织经验，分析图案结构、戗法运用、色线材料，然后制订复制方案，终于成功复制了的龙袍，作品获苏州市第二届民间艺术节金奖，自此，缂丝龙袍成为"王氏"批量生产的拳头产品和"专利"产品。《龙袍》参加第三届中国民间工艺品博览会暨第十届中国艺术博览会，再获金奖。2002年8月9日，中央电视台一套《夕阳红》节目播放了"王氏"五代缂丝世家的专题报道，"王氏缂丝"被越来越多的人所认识。

Wang Jialiang loves kossu fabrics more after retirement, and continues his career of weaving to explore and improve his skills. With his ancestral skills and his decades of practical experience in kossu fabrics, he gradually forms the artistic style and characteristics of "Wang's" kossu fabrics. First, he is good at weaving characters in which the style is unique and praised by the professionals in the field. Secondly, a new kossu product with three differences is created. Thirdly, he is good at weaving dragon robes with threads, which can be described as the traditional products of "Wang's family". Royal silk products such as dragon robes had already stopped working when Wang Jialiang was young, so that he had no perceptual knowledge of dragon robes. However, when experts from the Palace Museum found him, he decisively undertook the reproduction of the dragon robes. With many years of weaving experience, he analyzed the pattern, the application of the Qiang method and the use of colors and the materials of the threads, and then he worked out the plan for reproduction. Finally, he successfully reproduced an identical dragon robe, and his work won the gold medal in the second Suzhou Folk Art Festival. Since then, the kossu dragon robe has become the competitive product and "patented" product produced by Wang's quantity production. *Imperial Court Robe* participated in the Third China Folk Arts and Crafts Fair and the Tenth China Art Fair, and won the gold medal again. On August 9, 2002, CCTV's *Sunset Glow* program broadcast the special news of the five kossu generations of Wang Jialiang's family "Wang's kossu fabrics" is known by more and more people.

3. 成就与荣誉/Achievements and Honors

在1979年至1995年的16年中，王嘉良缂织的缂丝作品达几十幅，20世纪80年代初，凭借以多种技法缂织的《三星图》，王嘉良获得江苏省工艺美术精品"百花奖"。此后，写意精品《得意图》《金丝猴图》等作品相继面世。另有祖传题材及历史传统题材的《博古》《莲塘乳鸭图》《蝶穿牡丹》等，均为上乘之作，分别在市县范围内获奖。1990年3月，轻工业部为其颁发了"王嘉良同志从事工艺美术行业30年，为我国工艺美术事业的发展作出贡献"的荣誉证书。

During the 16 years from entering the factory in 1979 to retiring in 1995, Wang Jialiang woven dozens of kossu works. In the early 1980s, Wang Jialiang won the "Hundred Flower

Award" delivered by Jiangsu Arts and Crafts because of his work of *Three stars*. Since then, award–winning works such as *Elation Picture*, and *Golden Monkey Picture* have been published one after another. Other ancestral themes and historical and traditional themes, such as *Bogu*, *Suckling Ducks in the Lotus Pond* and "*Butterfly flying on the Peony*", are all excellent works, which won awards in cities and counties respectively. For this reason, he was rated as a craftsman in 1987. In March 1990, the Ministry of Light Industry awarded him the honorary certificate of "Mr. Wang Jialiang has been engaged in arts and crafts for 30 years and made contributions to the development of arts and crafts in China".

如果从乾隆年间算起，"王氏缂丝"应有200多年的历史；如果从同治年间的王新亭算起，也有150多年的历史。"王氏缂丝"的继承人非王嘉良莫属，现今王嘉良是苏州市民间工艺协会会员、江苏省民间工艺协会会员、江苏省高级工艺美术师，并享有"民间缂艺大师""民间工艺美术家"美誉。2008年5月，获"苏州市工艺美术大师"称号，2013年，被评为缂丝织造技艺省级非物质文化遗产代表性传承人。

Counted from the Qianlong period, the history of "Wang's kossu" should be more than 200 years. If counted from Wang Xinting in the Tongzhi period, it is also more than 150 years. However, the inheritor of "Wang's Kossu" is nobody except Wang Jialiang at present. Today, Wang Jialiang is a member of Suzhou Folk Arts and Crafts Association, a member of Jiangsu Folk Arts and Crafts Association, a senior arts and crafts artist in Jiangsu Province, and he enjoys the reputation of "Master of kossu Folk Arts" and the honorary title of "Folk Arts and Crafts Artist" in the city. In May 2008, he was awarded the title of "Master of Arts and Crafts in Suzhou", and the title of "Representative Inheritor of Provincial Intangible Cultural Heritage" in 2013.

二、王嘉良大师作品/Works of Master Wang Jialiang

王嘉良大师作品如图5-2-1～图5-2-3所示。他创作三异缂丝新品，代表作三异缂丝《寿桃图》，正面为寿桃图案，反面为篆书"寿"字，作品获2003年苏州市第二届民间艺术节金奖。

The Works of master Wang Jialiang are shown in Figures 5–2–1 to 5–2–3. He created a new product of kossu fabrics with three differences. The representative work of kossu fabrics *Longevity Peach* displays longevity peach pattern on the front and seal character "Shou" on the back. The work won the gold medal of the Second Suzhou Folk Art Festival in 2003.

三、王嘉良之子王建江/Wang Jianjiang, A Son of Wang Jialiang

1. 王建江简介/Introduction

（1）成长历程。王建江是苏州陆慕王氏缂丝世家的第六代传人。1976年王建江中学毕

图 5-2-1　缂丝《蝶穿牡丹》
Kossu *Butterflies Flying on the Peony*

图 5-2-2　缂丝《博古》
Kossu *Auspicious antifacts*

图 5-2-3　缂丝《寿桃图》
Kossu *Longevity Peach*

业后，父亲王嘉良便要求他跟随自己从事缂丝工作。那时的王建江还不能体会父亲的用心，也不明白为什么王家的男孩非要在织机前干活。尽管如此，孝顺的王建江还是跟着在德清、东台等地担任技术指导的父亲辗转于江浙一带，潜心学习缂丝技术。缂丝技艺易学难精，三年时间也只能掌握一些基本功。同时，艺人还必须具备一定的美术基础。"只有这样，才能看懂颜色的深浅变化。例如一朵牡丹花，只用三种颜色的丝线便没有层次感，十几种缂织出来的才好看。"在父亲的教导下，王建江逐渐与这门手艺熟络起来。

　　Personal development. Wang Jianjiang is the sixth generation descendant of Wang's kossu family in Lumu, Suzhou. After Wang Jianjiang graduated from the middle school in 1976, his father Wang Jialiang asked him to follow him in kossu fabrics. At that time, Wang Jianjiang couldn't understand his father, and he hardly realized why the boys in his family had to live and work in front of the loom. Despite this, Wang Jianjiang was filial and thus followed his father, a technical guide in Deqing and Dongtai, to work around Jiangsu and Zhejiang provinces to concentrate on the study of kossu fabrics technology. Kossu skills are easy to learn but difficult to master, and only some basic skills can be mastered in three years. What's more, a craftsman must have a certain artistic foundation. "Only in this way can we understand the change of colors. Like a peony flower, there is no changes brought about by only three colors of silk threads, and more than a dozen of threads can made it beautiful." Under the guidance of his father, Wang Jianjiang gradually became familiar with this craft.

　　1979年，父亲回到苏州，担任东山吴县缂丝总厂缂丝研究室副主任，王建江也成为厂里仅有的四名男学工之一。当时的入厂动机只是为了有份工作。随着技艺的精进，王建江在爱上了这门手艺的同时，也越来越折服于父辈祖辈坚守祖艺的精神。1983年他被调派到吴县陆慕缂丝厂技术科工作。1987年应邀到吴县陆慕张花缂丝厂工作，担任厂长和技术指导。1997年，王建江回到父亲组建的王氏缂丝世家工作室工作，与父亲合作了不少优秀作品。其中部分作品曾获得不同等级奖项。父子珠联璧合，也成为缂丝界的一段佳话。

In 1979, his father returned to Suzhou and served as the deputy director of the kossu Research Office of Dongshan Wu County Kossu General Factory. Wang Jianjiang became one of the only four male students in the factory. At that time, the purpose for entering the factory was just to have a job. With the improvement of his skills, Wang Jianjiang fell in love with this craft, and meanwhile, he was more and more impressed by his ancestors' spirit of sticking to ancestral art. In 1983, he was transferred to the technical department of Lumu Kossu Factory in Wu County. In 1987, he was invited to work in Lumu Zhanghua Kossu Factory in Wu County as the factory director and technical supervisor. In 1997, Wang Jianjiang returned to work in Wang's Kossu Family Studio set up by his father, and cooperated with his father to produce many excellent works, among which some works have won different kinds of prizes. The perfect cooperation between the father and the son has made a good story in the kossu industry.

（2）对缂丝的研究。王家在缂丝界声名在外，王嘉良受邀为故宫博物院修补龙袍。王建江得以随同父亲一睹宫廷缂丝精品的风采。2006年起，王建江先后为首都博物馆复制了元、明、清三代龙袍，其中元代"紫汤儿戏莲龙袍"由他一人独立完成。明万历皇帝衮服、清乾隆十二章龙袍，复制难度较大，需多位艺人分工协作。三件龙袍，前后用了七年时间。

The research on kossu fabrics. As the Wang family is well-known in the kossu world, Wang Jialiang was invited to repair the dragon robe for the Palace Museum. With his father, Wang Jianjiang was able to see the exquisite quality of the fine kossu of the royal court. Since 2006, Wang Jianjiang had reproduced the dragon robes of the Yuan, Ming and Qing Dynasties for the Capital Museum, among which the Dragon Robe "Children Play with Lotus in the Purple Pond" of the Yuan Dynasty was independently completed by himself. Ming Emperor Wanli's robes in the Ming Dynasty and dragon robes with twelve ornaments of the Qianlong period were difficult to reproduce, which required many craftsmen to work together. The three dragon robes took around seven years to complete.

2013年，故宫博物院院藏文物抢救性科技修复保护项目启动，王建江作为江苏唯一一位受邀的非物质文化遗产传承人参与了这次修复工作。修复讲究"修旧如旧"，先要把原材料做旧，再和文物上的丝线原色作对比，一丝一毫都马虎不得。只有确保彩线没有色差，才能继续织下去，有时某个衣角处都需要研究几天。尽管如此，王建江却甘之如饴，毕竟这代表了权威机构对王氏缂丝世家的认可。2015年12月成功为故宫博物院复制缂织攒竹嵌

168

玉石屏风，现陈列在故宫博物院寿康宫。2019年11月成功为故宫博物院复制寿康宫紫檀木嵌玻璃画七扇龙纹屏风。

In 2013, the salvaging technological restoration and protection project for cultural relics was launched in the Palace Museum. Wang Jianjiang, as the only invited inheritor of intangible cultural heritage of Jiangsu province, participated in the restoration work. Restoration pays attention to "repairing the old as the old". First, the raw materials should be made old, and then compared with the primary colors of silk threads on cultural relics carefully. Only by ensuring that there is no chromatic aberration in the colored threads can we continue to weave. Perhaps a certain corner of the clothes needs to be studied for several days. Despite this, Wang Jianjiang enjoyed it very much. After all, this represented the recognition of Wang's kossu family by authoritative organizations. In December of 2015, he successfully reproduced the kossu bamboo and jade screen for the Palace Museum, which has been displayed in Shoukang Palace of the Palace Museum. In November of 2019, seven dragon-patterned ornament screens of Shoukang Palace's purple inlaid glass paintings were successfully reproduced for the Palace Museum.

20世纪80年代前后，由于日本和服腰带上有缂丝用料的需求，曾使缂丝一度走俏。不少人纷纷跟风做起缂丝生意，导致市场迅速饱和。至20世纪90年代，缂丝需求急剧萎缩，从业者又大量转行。由于人工成本大，价格较高，又无法用机器量产，因此缂丝市场一直很难得到根本性的开拓。但王建江没有放弃传承与创新。近年来，他将目光瞄准工艺品市场，生产紧跟现代市场的产品，制作缂丝团扇、山水画屏风、围巾、钱包、香囊、宫扇等精致的小饰品。并在材质上积极寻求创新，灵活采用羊毛等材料，即"缂毛"。此外，他还接受高端定制，为一些品牌服装提供精美的缂丝装饰。

Around the 1980s, due to the demand for kossu materials on Japanese kimono belts, kossu fabrics were once popular. Many people had followed the trend to start the kossu business, which led to the rapid saturation of the market. By the 1990s, the demand for kossu fabrics shrank sharply, and a large number of practitioners switched careers. Because of the high labor cost, the high price and the inability to mass-produce with machines, it had been difficult to develop the kossu market fundamentally. But Wang Jianjiang did not give up inheritance and innovation. In recent years, he has aimed at the handicraft market, produced products closely following the modern market, and made exquisite trinkets such as round kossu fans, landscape kossu painting screens, scarves, wallets, sachets and mandarin fans. He actively seeks for innovation in materials, flexible use of wool and other materials, that is, "wool". In addition, he also accepts high-end customization and provides exquisite kossu decoration for some high-end clothing brands.

在王建江看来，缂丝经纬交织，若将经线比作人生的长度，那么色彩斑斓的纬线则像极了丰富多彩的人生宽度。他也凭借《龙袍》《正龙》等作品获奖颇丰，现已成为非物质文化遗产代表性传承人。作为王氏第六代传人，如何将这门逐渐冷清的技艺传承下去，已成为他最关心的问题。王建江说，无论外面的世界向何方发展，作为王氏后人，都有责任将这门技艺传承下去。

In Wang Jianjiang's view, the warp and weft of the kossu fabrics are intertwined. If the warp is compared to the length of life, the colorful weft is like the rich width of life. He has also won many awards for his works such as *The Dragon Robe* and *Dragons*, and has now become the representative inheritor of intangible cultural heritage. As the sixth generation descendant of Wang's family, how to pass on this gradually deserted skill has become his greatest concern. Wang Jianjiang says that no matter where the outside world will develop, as descendants of Wang's family, they have the responsibility to pass on this skill.

2. 成就与荣誉/Achievements and Honors

王建江，苏州市非物质文化遗产代表性传承人，技能大师，高级工艺美术师，江苏省乡土人才"三带"名人，姑苏高技能重点人才。

2006年，和父亲王嘉良合作的缂丝作品《龙袍》荣获第十三届中国艺术博览会金奖。

2009年，参加工艺美术大展暨庆祝中华人民共和国成立60周年并获奖，同年参加苏州市首届旅游产品商品评选并获优秀奖。

2010年8月，《龙袍》荣获第五届中国民间工艺品博览会金奖。

2010年9月，《正龙》荣获"艺博杯"江苏省工艺美术精品大奖赛银奖。

2011年，《龙袍》荣获"艺博杯"江苏省工艺美术精品大奖赛金奖。

Wang Jianjiang is the representative inheritors of the intangible cultural heritage of Jiangsu City, a skilled master, a senior arts and crafts artist, a celebrity of "Three Zones" of local talents in Jiangsu Province, and a key high-skilled talent in Suzhou.

In 2006, the kossu work *The Dragon Robe*, which was done by cooperation with his father, won the gold medal in the 13th China Art Fair.

In 2009, he participated in the Arts and Crafts Exhibition and celebrated the 60th anniversary of the founding of the People's Republic of China. In the same year, he participated in the first tourism product selection in Suzhou and won the excellence award.

In August 2010, *The Dragon Robe* won the gold medal in the 5th China Folk Arts and Crafts Expo.

In September 2010, *Dragons* won the silver medal in "Art Cup" Arts and Crafts Grand Prix in Jiangsu.

In 2011, *The Dragon Robe* won the gold medal in "Art Cup" Arts and Crafts Grand Prix in Jiangsu.

◎ 思考题/Questions for Discussion

1．王嘉良哪个作品获得苏州市"四新"产品二等奖？/Which work of Wang Jialiang won the second prize of "Four New" products in Suzhou?

2．"王氏缂丝"艺术风格和特色有哪些？/What are the artistic styles and characteristics of Wang's kossu fabrics?

3．王嘉良大师作品中哪些是历史题材？/What are the historical themes in Master Wang Jialiang's works?

4．王建江独立完成的龙袍复制叫什么？/What is the name of Wang Jianjiang's independent reproduction of the dragon robe?

第三节　马惠娟/Ma Huijuan

一、马惠娟大师简介/Introduction to Master Ma Huijuan

马惠娟大师

1．成长历程/Personal Development

马惠娟，1953年生，胥口蒋墩人，因家庭贫困，只读了两年小学便随母亲学习刺绣。因聪明好学，少女时期的马惠娟便掌握了民间流行针法，能独立完成各类绣品，成为家庭经济收入的主要成员。成年后，经蒋雪英推荐，马惠娟进入吴县机绣厂。时值吴县抢救缂丝古艺，成立缂丝车间（即后来的吴县缂丝总厂），从蠡口等地请来6名缂丝老艺人传授缂丝技艺，马惠娟有幸成为该厂第一批艺徒，师承老艺人沈根娣学习缂丝。

Ma Huijuan, born in 1953 in Jiangdun of Xukou, studied embroidery with her mother after only two years of primary school because of poverty. Because she was smart and eager to learn, Ma Huijuan mastered the popular folk stitches in her girlhood, and was able to complete all kinds of embroidery independently. And thus, she became the major member of the bread-earner of her family. When she grew up, Ma Huijuan, recommended by Jiang Xueying, entered Wu County Machine Embroidery Factory. At that time, Wu County rescued the ancient kossu techniques, set up a kossu workshop (later the Wu County Kossu General Factory), and invited six elder kossu craftsmen from Likou and other places to teach kossu techniques. Ma Huijuan was fortunate to be the first batch of apprentices in the factory, and studied kossu techniques under the elder craftsman Shen Gendi.

当时厂里承接了缂制和服腰带的业务，马惠娟等6名艺徒面临着继承掌握传统戗法技艺以及与西洋文化相结合的双重难题，马惠娟在老艺人的帮教下虚心学、勤奋练，夜以继日、废寝忘食，经过6个月的努力，终于试织成了第一条缂丝腰带，当日本亚东国际贸易公司的阪下先生见到后欣喜若狂，惊呼"中国传统的缂丝工艺依然存在"。腰带的缂织成功，不仅打开了缂丝产品通往日本的销路，而且创造了可观的经济效益，更为重要的是救活了县境有千年历史的传统行业，并为其兴旺发展打下了基础。在当时，一条300工时以上的缂丝腰带，可换回27吨钢材、120辆自行车或一辆丰田轿车，而马惠娟一年可缂织2条450～600工时的高难度腰带，不仅工时超倍，而且质量名列前茅。因此，不到一年，马惠娟就成为带徒授艺的小师傅。1973～1987年，只吴县缂丝总厂一家就出口缂丝腰带7400多条，创汇达1千多万元人民币。

When China and Japan resumed diplomatic relations, the factory undertook the business of weaving kossu kimono belts. Ma Huijuan and other six craftsmen were faced with the dilemma of inheriting and mastering traditional Qiang weaving and combining it with Western culture. With the help of the elder craftsman, Ma Huijuan studied modestly, practiced diligently, stayed up late and worked day and night. After six months' efforts, he finally tried to finish the first kossu belt. When Mr. Sakashita of Japan Yadong International Trading Company saw it, he was ecstatic and exclaimed "the traditional Chinese kossu techniques still exists". The successful weaving of kossu belts not only opened up the market of kossu products to Japan and created considerable economic benefits, but also saved the traditional industries with a history of thousands of years in the county and laid a foundation for their prosperous development. At that time, a kossu belt with more than 300 working hours could be exchanged for 27 tons of steel, 120 bicycles or a Toyota car, while Ma Huijuan could weave two belts overcoming difficulties with 450 to 600 working hours a year, which not only doubled the working hours, but also ranked among the best in quality. Therefore, in less than a year, Ma Huijuan became superfine master who taught art with disciples. In the 15 years from 1973 to 1987, Wu County Kossu Factory exported more than 7,400 kossu belts and earned more than 10 million yuan in foreign exchange.

1984年，吴县缂丝总厂成立缂丝研究所，下设精品车间，马惠娟成为精品研制的专业人员。马惠娟的精湛技艺和精美作品受到了日商的关注。同年5月，她应日本亚东株式会社之邀东渡日本，在京都、熊本、大分、大阪等9个城市作了9场缂丝表演，所到之处受到日本人民的欢迎和礼遇。她缂织的作品精细、挺括，图案形象逼真，花草有生机活力，流水有动感，动物呼之欲出，人物达到摄像的效果。百元成本的生丝底料和色线，经她精心缂织后便身价百倍，一幅《泼墨乌龙》缂丝，外贸收购价达6万元人民币，进入日本市场后还可翻上几倍。随之而起的是日本市场经久不衰的中国缂丝腰带的销售热潮。

In 1984, Wu County Kossu General Factory established Kossu Research Institute with a boutique workshop, and Ma Huijuan became a professional in boutique development. Ma Huijuan's superb skills and exquisite works attracted the attention of Japanese businessmen. In

May of the same year, she traveled to Japan at the invitation of Japan Yadong Co., Ltd., and gave nine kossu weaving performances in nine cities including Kyoto, Kumamoto, Oita and Osaka, where she was welcomed and treated with courtesy by the Japanese people. The kossu works she weaved were fine and crisp, with vivid patterns, vigorous flowers and plants, dynamic running water, animals ready to come out, and characters achieved the effect of photography. The raw silk, base materials and color threads with a cost of 100 yuan were worth 100 times after being carefully woven by her. Splashing Ink Oolong, a piece of kossu picture, had a foreign purchase price of 60,000 yuan, which could be doubled several times after entering the Japanese market. What followed was the enduring sales boom of Chinese kossu belts in the Japanese market.

1986年，马惠娟任吴县缂丝学会理事。1987年被评为工艺技师，并被吸收为中国工艺美术协会会员。

In 1986, Ma Huijuan was appointed as the director of Wu County Kossu Council. In 1987, she was rated as a craftsman and absorbed as a member of China Arts and Crafts Association.

2. 对缂丝的研究/Researches on Kossu Fabrics

1980年，吴县缂丝厂老艺人陈阿多（县政协委员）随中国工艺美术代表团去中国香港作缂丝表演，引起强烈反响。翌年，便有香港霍英东之女霍丽娜小姐慕名来厂，要求复制一幅大型缂丝《莲塘乳鸭图》，这是宋代吴县缂丝名匠朱克柔的代表作，原作藏上海博物馆，马惠娟拿到的仅是一张5英寸的照片，在陈阿多老师傅的帮助和鼓励下，她克服重重困难，用时3个多月，终于攻克了历史名作的复制难关。当霍小姐见到这幅复制品时，惊喜地赞叹道："工之精，可乱真。"历史名作的复制过程，也是熟悉掌握传统技艺的过程，马惠娟从中学到了许多传统戗法。1982年，马惠娟承接"抽拉雕绣"腰带新品开发任务，在研制过程中，她既会缂丝又会刺绣的优势得到了发挥，经反复试验，不断改进，终于达到了预期艺术效果，日商大批量订货，只一个产品就做了3年，合计产值达600多万元。

In 1980, Chen Aduo, an elder craftsman from the Wu County Kossu Factory (member of the County Political Consultative Conference), went to Hong Kong with the Chinese arts and crafts delegation to perform kossu weaving, which aroused strong repercussions. In the following year, Miss Huo Lina, the daughter of Huo Ying–tung from Hong Kong, came to the factory to request a copy a large–scale kossu work of Suckling Ducks in the Lotus Pond, which was the representative work of Kerou Zhu, a famous kossu craftsman in Wu County in the Song Dynasty. The original work is collected in Shanghai Museum, and Ma Huijuan only got a 5–inch photo. With the help and encouragement of Master Chen Aduo, she overcame many difficulties and spent more than three months finally completing copying historical masterpieces. When Miss Huo saw this replica, she was amazed and praised "The expertise of the craft can make it real." The copying process of historical masterpieces is also a process of familiarizing yourself with traditional techniques, from which Ma Huijuan learned many traditional methods. In 1982, Ma Huijuan undertook the task of developing a new belt of "drawing, carving and embroidering". During the development process, her advantages of both kossu weaving and embroidery were brought into full play. After repeated trials and continuous improvement, the expected artistic effect was finally achieved. Japanese businessmen ordered in large quantities, and only one product was produced for three years, with a total output value of more than 6 million yuan.

马惠娟熟谙历代技法，结合织造实践开展理论探索，多篇论文被专业刊物录用，在缂丝图案题材、色线等方面多次创新。然而，吴县缂丝业欲重振雄风，进一步走向世界，必须研制高品位的艺术精品，必须在题材、底料、色线以及戗法技艺上有所创新。马惠娟在陈阿多等老艺人的帮助下开始了缂丝艺术精品的试制，并很快成长为艺术欣赏品的缂织高手。

Ma Huijuan is familiar with the techniques of past dynasties, and carries out theoretical exploration in combination with weaving practice. Many papers have been accepted by professional journals, and she has made many innovations in the theme of Kossu patterns and color lines. However, in order to revive the kossu industry in Wu County and go further to the world, it is necessary to develop high-grade artworks, and to innovate in subject matter, base materials, color threads and techniques. With the help of Chen Aduo and other elder craftsmen, Ma Huijuan started the trial production of kossu artworks, and quickly grew into a master of kossu weaving for artistic appreciation.

在缂丝研究所设计师马超骏的合作下，她一次又一次地攻克技术难关，试制了一系列艺术精品。自以五彩羊毛为材料缂织《献寿图》获得成功后，又借鉴刺绣技艺试缂了《虎啸图》。虎，是缂丝行业从未触及过的披毛动物题材，马惠娟借鉴了刺绣技艺中散套针法并运用到缂丝传统的长短戗、斜戗之中，使纬线呈放射状伸展，增强了虎毛的质感，解决了缂织披毛动物的一大难题。

Through the cooperation with Ma Chaojun, a designer of Kossu Research Institute, she has overcome technical difficulties again and again and trial-produced a series of artworks. Since the success of weaving the kossu work of *Birthday Picture* with colorful wool as the materials, she tried to weave the kossu work of *Roaring Tiger* with reference to embroidery skills. Tiger is the theme of hairy animals that has never been touched by the kossu industry. Ma Huijuan borrowed the scattered trocar method in embroidery skills and applied it to the traditional long and short Qiang weaving and oblique Qiang weaving, which made the weft stretch radially, enhanced the texture of tiger hair and solved a big problem of weaving kossu fabrics of hairy animals.

马惠娟之所以能成为中国缂丝行业顶尖高手，除了她在工作上认真负责，技艺上精益求精，还几乎为缂丝业献上了所有的业余时间。为了提高自己的艺术修养，她自费参加了上海东方艺校的绘画函授，她克服文化低的困难攻读各种理论研究文章，研究总结传统缂丝技法，并结合自己的实践经验和心得体会，写出了《缂丝传统技法的运用》《缂丝戗法与现代缂丝》等4篇学术论文，分别发表于专业杂志。

The reasons why Ma Huijuan became the top master of China's kossu weaving industry were that she was serious and responsible in her work and kept improving her skills, and almost gave up all her spare time. In order to improve her artistic accomplishment, she attended the painting correspondence course of Shanghai Oriental Art School at her own expense. Overcoming the difficulty of lack of education, she studied various theoretical research articles, studied and summarized traditional kossu techniques. Combined with her own practical experience and experience, she wrote four academic papers, such as *Application of Traditional Kossu Techniques*, *Kossu Weaving Method and Modern Kossu Techniques*, which were published in professional magazines respectively.

除了她自身的努力和设计技师马超骏的合作帮助外，她的成功更离不开前辈老艺人的帮教。除了启蒙师傅沈根娣之外，她还得到陈阿多与徐祥山的传授。马惠娟为缂丝技艺的继承发展，为企业的经济效益，为家乡的经济发展、出口创汇所做出的贡献，得到了社会的认可。1982年，马惠娟被中国工艺美术学会吸收为会员，1993年被破格批准为国家级缂丝技师（当时全国仅有4人获此职称），之后又先后获得了苏州市"十佳职工"、江苏省"最佳主人翁"称号。

Ma Huijuan's success not only due to her own efforts and the cooperation and help of designer Ma Chaojun, but also due to the help and teaching of senior craftsmen. In addition to Shen Gendi, the enlightenment master, she also benefited from the teaching of Chen Aduo and Xu Xiangshan. Ma Huijuan's contribution to the inheritance and development of kossu techniques, the considerable economic benefits for enterprises, the economic development of his hometown and the earning of foreign exchange through export, have been recognized by the society. In 1982, Ma Huijuan was granted as a member by China Arts and Crafts Council, and in 1993, he was exceptionally approved as a national kossu technician (at that time, only four people won this title). Later, she was awarded the honorary titles of "Top Ten Employees" in Suzhou City and "Best Master" in Jiangsu province.

20世纪90年代后，马惠娟在缂丝技艺上又有新的创举，她以摄影照片为底稿，在她的木制缂机上织成了日本东大教授的人像，这是国内第一幅以人像为主题的缂丝精品，东大教授深知缂丝古艺缂织人物肖像的难度，他本人只抱有"三分像"的希望和要求，岂料马惠娟缂织的《东大教授像》比照片更精神。缂丝人像的成功，再一次拓宽了缂丝的图案题材，江苏省电视台适时为缂丝业的创新做了报道，作品在苏州刺绣研究所展出时得到了业内专家的高度评价。马惠娟于1997年吴县缂丝总厂改制时离开企业，2003年退休。退休后，她在缂丝技艺和题材方面又有新的突破，缂织了抽象画《水乡古镇》等几十幅艺术精品，具有极高的艺术欣赏价值。

After the 1990s, Ma Huijuan has made new innovations in kossu techniques. She used photographs as the manuscript and woven a portrait of a professor at Tokyo University on her wooden loom. In China, this is the first fine kossu fabrics using portrait as the theme. The Professor of Tokyo University was well aware of the difficulty of weaving portraits with ancient kossu techniques, and he only had the hope and requirement of 30% resemblance. Unexpectedly, the portrait of the Professor woven by Ma Huijuan was more vivid than photos. The success of the kossu portrait once again broadened the pattern theme of the kossu fabrics. The provincial TV station reported on the innovation of the kossu industry, and the works were highly praised by experts in the industry when they were exhibited in Suzhou Embroidery Research Institute. Ma Huijuan left the enterprise in 1997 when Wu County Kossu General Factory was restructured, and she retired in 2003. After her retirement, she made new breakthroughs in kossu techniques and the theme of the kossu fabrics. She has woven abstract painting *Ancient Town of Waterland*, which have high artistic appreciation value.

3. 成就与荣誉/Achievements and Honors

马惠娟为苏州市缂丝非物质文化遗产代表性传承人。1984年，她缂织的《虎啸图》被选送参展北京亚太地区国际博览会，获得银奖。1984年，马惠娟应日本亚东国际贸易公司邀请赴日本作缂丝表演，其作品多次在省、全国的工艺品评比中获奖，并4次选送国外展出。1985年试制的《孔雀羽腰带》和《古寒山寺》缂丝屏，参加全国同行业评比获总分第一名，并获第六届中国工艺美术百花奖金杯奖。之后又先后研制了《寒山夕照》《博古》等缂丝新品和精品；《寒山夕照》参加第二届北京国际博览会获银奖，《博古》获全国缂丝研究会中青年艺人作品评比二等奖。1986年缂织《寒山寺》《孔雀羽腰带》，在第六届中国工艺美术百花奖评比中总分第一，为吴县缂丝总厂产品获金杯奖做出贡献。2006年10月，作品《徐悲鸿六骏图》获苏州工艺美术精品展金奖。2007年被评为研究员级高级工艺美术师，并获苏州市"第二届苏州民间工艺家"称号。2007年11月，马惠娟的《寒月孤雁》缂丝精品参展杭州第八届中国工艺美术大师作品暨工艺美术精品博览会，再获金奖。2008年苏州市人民政府授予其"工艺美术大师"的称号。

Ma Huijuan is the representative inheritor of kossu intangible cultural heritage in Suzhou. In 1984，her *Roaring Tiger* was selected to participate in Beijing Asia-Pacific International Expo and won the silver prize. In 1984, at the invitation of Japan Yadong International Trade Company, Ma Huijuan went to Japan to perform kossu weaving. His works won many awards in provincial and national handicraft competitions, and they were sent to foreign countries for exhibition four times. The *Peacock Feather Belt* and the kossu screen of *Ancient Hanshan Temple*, which were trial–produced in 1985, won the first place in the total score in the national industry appraisal and won the Golden Cup Award of the 6th China Arts and Crafts Hundred Flowers Award. After that, she has successively developed new kossu products and fine products such as *Hanshan Sunset* and *Bogu*. *Hanshan Sunset* won the silver award in the Second Beijing International Expo, and *Bogu* won the second prize in the appraisal of young and middle–aged artists' works by the National Kossu Research Association. In 1986, *Hanshan Temple* and *Peacock Feather Belt* were woven, which won the first total score in the 6th China Arts and Crafts Hundred Flowers Award, and contributed to the Golden Cup Award for the products of Wu County Kossu General Factory. In October 2006, the work *Xu Beihong's Six Horses* won the gold medal in the Fine Arts and Crafts Exhibition in Suzhou. In 2007, he was approved as a research–level senior arts and crafts artist and won the title of "the second Suzhou folk craftsman" in Suzhou. In November 2007, Ma Huijuan's *Lonely Goose in Cold Moon* was exhibited in the 8th China Arts and Crafts Master Works and Arts and Crafts Boutique Expo in Hangzhou, and won the gold medal again. In 2008, Suzhou Municipal People's Government awarded her the title of "Master of Arts and Crafts".

二、马惠娟大师作品/Works of Master Ma Huijuan

《春溪浴鸭图》（图5-3-1）纵175cm，横75cm，马惠娟摹缂，原作为清宫所制缂丝名作。

Bathing Ducks in the Spring Creek (Figure 5–3–1) reproduced by Ma Huijuan, is 175 cm high and 75 cm wide, and the original work, as a famous kossu work was made by the Qing Palace.

《崔白三秋图》（图5-3-2）纵105cm，横45cm，马惠娟摹缂，原作为清宫所藏明代缂丝名作。

The Autumn for Cui Bai (Figure 5–3–2) is 105 cm in length and 45 cm in width, reproduced by Ma Huijuan, and the original work, as a famous kossu work of the Ming Dynasty collected by the Qing Palace.

《斗雀》（图5-3-3）纵61cm，横32cm，陈之佛原画，马惠娟摹缂。

Fighting Birds (Figure 5–3–3) is 61 cm in length and 32 cm in width. The original work was painted by Chen Zhifo and it was copied with kossu by Ma Huijuan.

此外还有《老子青牛》（图5-3-4）和《于非闇拟宋缂丝》（图5-3-5）。

There are also *Lao Zi's Bull Riding* (Figure 5–3–4) and *Yu Feian's Imitation of Kossu Fabrics in the Song Dynasty* (Figure 5–3–5).

图5-3-1　缂丝《春溪浴鸭图》
Kossu *Bathing Ducks in the Spring Creek*

图5-3-2　缂丝《崔白三秋图》
Kossu *The Autumn for Cui Bai*

图5-3-3　缂丝《斗雀》
Kossu *Fighting Birds*

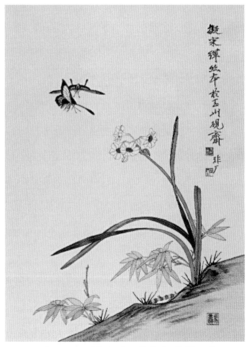

图5-3-4　缂丝《老子青牛》
Kossu *Lao Zi's Bull Riding*

图5-3-5　缂丝《于非闇拟宋缂丝》
Kossu *Yu Feian's Imitation of Kossu Fabrics in the Song Dynasty*

◎ 思考题/Questions for Discussion

1. 1984年5月，马惠娟精湛技艺和精美作品引起日商关注，哪一幅缂丝作品以六万人民币的价格进入日本市场？/In May 1984, Ma Huijuan's exquisite skills and works attracted the attention of Japanese businessmen. Which Kossu work was sold in the Japanese market at a price of 60,000 RMB?

2. 1980年，哪位缂织大师随中国工艺美术代表团去香港进行缂丝表演？/ In 1980, who went to Hong Kong with the Chinese Arts and Crafts Delegation to perform kossu weaving?

3. 宋代吴县缂丝名将朱克柔的代表作是什么？/What is the representative work of Zhu Kerou, a famous silk reeling star in Wu County in the Song Dynasty?

第四节　范玉明/Fan Yuming

一、范玉明大师简介/Introduction to Master Fan Yuming

范玉明大师

1. 成长历程/Personal Development

范玉明，1973年出生于苏州市东渚镇，在太湖边长大，从小接受家庭文化的熏陶，虽然生活艰苦，但仍旧认真读书。其母擅长刺绣、缂丝，他耳濡目染，于是将刺绣、缂丝等织造技艺熟记于心。长大后，他对缂丝情有独钟，拜王金山大师为师，励志在缂丝领域有所造诣、殚精竭虑、勤勤恳恳，跟随着王金山大师研究缂丝。经二十个春夏秋冬的研究，他在花鸟虫鱼、佛道人物、山水建筑等图案方面不断创新，其缂丝技艺日趋成熟，独自成为一派。

Fan Yuming, born in Dongzhu, Suzhou in 1973, grew up on the shore of Taihu Lake, and was trained by the court. His mother was good at embroidery and kossu weaving. He often witnessed and learned the traditional skills from his mother. When he grew up, he focused on kossu fabrics, and he studied under the guidance of Master Wang Jinshan, determined to try his best to follow the master to study kossu techniques. For more than 20 years, he constantly made innovations on ancient kossu techniques in terms of weaving flowers, birds, insects and fishes, Buddhist and Taoist figures, landscapes and etc. As his kossu techniques became mature, he formed his own weaving style.

2. 对缂丝的研究/Researches on Kossu Fabrics

2003年，范玉明创立公司，从事缂丝技艺的研习、传承。2006年，范玉明受明十三陵博物馆邀请与南京云锦研究所数位专家一同参与出土文物复制项目，通过描摹原物，分析研究织物组织结构，参考文献古籍，根据发掘报告还原色彩，最终形成设计图样。通过数年努力，成功复制"明黄地万历十二章福寿衮服""寿桃毡靴"等十余件文物。2011年，范玉明被山东丝绸纺织职业学院（现山东轻工职业学院）聘为客座教授，传授缂丝理论和操作技艺。同年，范玉明进入清华大学美术学院进修纤维专业，跟随导师林乐成教授进行纤维艺术研究。2015年，范玉明师从王金山大师，创立"王金山大师缂丝传艺班"。

He was engaged in the study and inheritance of kossu techniques, and founded his company in 2003. In 2006, Fan Yuming was invited by Ming Tombs Museum to participate in the reproduction project of unearthed cultural relics together with several experts from Nanjing Yun Brocade Research Institute. He analyzed and studied the structure of the fabrics by tracing the original objects, referring to ancient books, and restoring the color according to the excavation report. After several years of hard work, more than ten cultural relics have been successfully copied, such as "Ming Huangdi Wanli's Robe with Twelve Patterns Good Fortune" "Boots with the Design of Longevity Peach". In 2011, he was employed as a visiting professor by Shandong Silk Textile Vocational and Technical College (How Shandong Vocational College of Light Industry) to teach kossu weaving theory and operation skills. In the same year, he entered the Acadeing of Arts and Design of Tsinghua University to study in fiber major, and followed his adviser Professor Lin Lecheng to study fiber art. In 2015, he studied under Master Wang Jinshan and founded the "Master Wang Jinshan Kossu Art Class".

为进一步传播非遗文化，传授缂丝技艺，范玉明与苏州旅游与财经高等职业技术学校、苏州工艺美术职业技术学院等多所学校进行长期合作，创办了"范玉明缂丝织造技艺大师工作室"，几年时间里，培养了多名学生，使缂丝这门古老的技艺薪火相传，生生不息。工作室被评为"江苏省乡土人才大师工作室"和"苏州市技能大师（名师）工作室"。

In order to further spread the intangible culture and teach the kossu techniques, Fan Yuming has long–term cooperation with Suzhou Tourism and Finance Higher Vocational Technical School, Suzhou Art and Design Technology Institute and many other schools, and established the "Master Studio of Fan Yuming Kossu Weaving Skills". In the past few years, many students have been trained, so that the ancient kossu techniques can be passed down from generation to generation. The studio was rated as "Master Studio of Jiangsu Local Talent Skills" and "Suzhou Skill Master (Famous Teacher) Studio".

3.成就与荣誉/Achievements and Honors

范玉明创作的《和谐社会》获首届中国当代纤维艺术展传统技艺奖，缂丝作品《风骨》

获得了"从洛桑到北京"第七届国际纤维艺术双年展优秀奖。2019年，范玉明前往中国艺术研究院跟随邱春林导师学习访问，期间创作的《墨牡丹》参加首届国家博物馆工艺美术作品邀请展，并被中国工艺美术馆永久收藏。

The Harmonious Society created by Fan Yuming won the traditional skill award in the first China Contemporary Fiber Art Exhibition, and the kossu work *Vigour of Style* won the excellence award at the 7th "From Lausanne to Beijing" International Fiber Art Biennale and Symposium. In 2019, he went to the China Academy of Arts to follow the advisor Qiu Chunlin to study the art courses for visiting scholar. During this period, *Ink Peony* created during the period participated in the lst Invitational Exhibition of Arts and Crafts held in National Museum of China and was permanently collected by China National Museum of Arts and Crafts.

范玉明为研究员级高级工艺美术师、苏州市缂丝非物质文化遗产代表性传承人、江苏省名人、苏州市高级工艺美术师、中国织锦工艺大师、中国名族工艺美术大师、苏州市工艺美术学会缂丝专业委员会副会长、苏州市民间文艺家协会副秘书长。

Fan Yuming is a researcher–level senior arts and crafts artist, a representative inheritor of Suzhou Kossu Intangible Cultural Heritage, a celebrity in Jiangsu Province, a senior arts and crafts artist in Suzhou, a master of Chinese brocade craftsmanship, a master of Chinese famous arts and crafts, the vice president of Kossu Professional Committee of Suzhou Arts and Crafts Council, and deputy secretary general of Suzhou Civil Writers Association.

二、范玉明大师作品/Works of Master Fan Yuming

《六臂勇保护法》(图5-4-1)在2009年获得中国工艺美术"百花奖"(深圳)优秀作品银奖。《十二章纹九龙袍服》(图5-4-2)获得2009年中国南京文化产业交易会·江苏艺博杯工艺美术精品奖银奖。

In 2009, *Six-Armed Protection Buddha* (Figure 5-4-1) won the silver award for Outstanding Works of Chinese Arts and Crafts "Hundred Flowers Award" (Shenzhen). *Nine-Dragon Robe with Twelve Patterns* (Figure 5-4-2) won the silver award of 2009 China Nanjing Cultural Industry Fair—Jiangsu Art Expo Cup Arts and Crafts Excellence Award.

《荷花图》(图5-4-3)获得2011中国(深圳)国际文化产业博览交易会"中国工艺美术文化创意奖"银奖。《春溪水族图》(图5-4-4)获2017中国(青岛)工艺美术博览会"金凤凰"创新设计大赛金奖。

Lotus (Figure 5-4-3) won the silver award of "Chinese Arts and Crafts Cultural Creativity Award" at the 2011 China (Shenzhen) International Cultural Industry Expo. *Aquatics in the Spring Creek* (Figure 5-4-4) won the gold award of "Golden Phoenix" at the 2017 China (Qingdao) Arts and Crafts Expo.

图5-4-1　缂丝《六臂勇保护法》
Kossu *Six-Armed Protection Buddha*

图5-4-2　缂丝十二章纹九龙袍服
Kossu *Nine-Dragon Robe with Twelve patterns*

图5-4-3　缂丝《荷花图》
Kossu *Lotus*

图5-4-4　缂丝《春溪水族图》
Kossu *Aquatics in the Spring Creek*

◎ 思考题/Questions for Discussion

1. 范玉明师从哪位大师？/Which master did Fan Yuming learn from?

2. 为进一步传播非遗文化，传授缂丝技艺，范玉明与多所学校进行长期合作，创办了什么？/In order to further spread the intangible culture and teach the kossu techniques, Fan

Yuming cooperated with many schools for a long time. What did he establish?

3．范玉明作品《春溪水族图》获 2017 中国（青岛）工艺美术博览会什么奖项？／What award did Fan Yuming's work Aquatics in the Spring Creek win at the 2017 China(Qingdao)Arts and Crafts Expo?

4．获得"2009 年中国南京文化产业交易会·江苏艺博杯工艺美术精品奖"的是什么作品？／What works won the "2009 China Nanjing Cultural Industry Fair—Jiangsu Art Expo Cup Arts and Crafts Excelence Award"？

第五节　蔡霞明/Cai Xiaming

一、蔡霞明大师简介/Introduction to Master Cai Xiaming

1．成长历程/Personal Development

蔡霞明大师

蔡霞明，1968 年出生于苏州市吴中区，1986 年在苏州市长桥缂丝厂跟随缂丝前辈李才福、莫忠英学缂丝技艺，截至目前从事缂丝织造已有 34 年。1986 年至 1997 年，在苏州市长桥缂丝厂担任质检科长。1998 年至 2005 年，在苏州鹏程工艺服饰有限公司担任生产技术厂长并招收学徒进行技术培养，对公司的发展和产品创新起到了决定性作用，受日方邀请，多次参加技术交流及产品展示会的现场表演。2007 年与苏州禾孚泰贸易有限公司合作缂丝、刺绣培训及教学工作。2009 年至今，创办了苏州市吴中区长桥蔡氏绣庄，并担任总设计师。2016 年，被吴中区临湖成艺工艺品厂聘请为技术顾问，负责技术指导、配色制作等工艺。2016 年 12 月，被江苏省人力资源和社会保障厅评为高级工艺美术师。2017 年至今，创办了苏州蔡霞明织绣品有限公司，并担任总设计师。

Cai Xiaming was born in Wuzhong District, Suzhou City in 1968. In 1986, he followed the Kossu predecessors Li Caifu and Mo Zhongying to learn kossu techniques in Suzhou Changqiao Kossu Factory. Up to now, he has been engaged in kossu craftsmanship for 34 years. From 1986 to 1997, he served as the head of quality inspection in Suzhou Changqiao Kossu Factory. From 1998 to 2005, he worked as the production technology director of Suzhou Pengcheng Craft Clothing Co., Ltd. and recruited apprentices for technical training, which played a decisive role in the development and product innovation of Suzhou Pengcheng Craft Clothing Co., Ltd. At the invitation of the Japanese side, he participated in technical exchanges and live performances of product exhibitions for many times. In 2007, he cooperated with Suzhou Hefutai Trading Co., Ltd. in kossu and embroidery training and teaching. From 2009 to now, he founded Changqiao Cai's

Embroidery Village in Wuzhong District, Suzhou City, as the chief designer. In 2016, he was employed as a technical consultant by Linhu Chengyi Arts and Crafts Factory in Wuzhong District, responsible for technical guidance, color matching and other processes. In December 2016, he was rated as a senior arts and crafts artist by Jiangsu Provincial Department of Human Resources and Social Security. From 2017 to now, he founded Suzhou Cai Xiaming Weaving Embroidery Co., Ltd., and served as the chief designer.

2. 对缂丝的研究/Researches on Kossu Fabrics

蔡霞明不仅掌握了基本的缂丝传统技艺，还掌握了缂丝中的两大技术——明缂丝和本缂丝，包括材质要求与规格、描稿、配色、扦经、接经、打番头、刷经面、贴经面、摇线、缂织、剪毛头等整个技艺。此外，她开创了新的作品艺术表现形式，例如，人物肖像《玛丽莲·梦露》，是通过将人物图像转变成漫画的形式，再用缂丝的方法织显出来；在缂织中，将刺绣的相色法运用到缂丝上，例如，油画颜色众多的缂织，打破了原始的单色、两色相戗法，在丰富的颜色上运用三色、四色相拼，通过人为捻线的方法，根据不同的捻度产生不同的效果进行缂织。她通过研究试作，成功地缂织出王金山大师之后无人掌握的两面三异技艺作品。2018年，蔡霞明团队根据马王堆出土的金丝羽衣，开发了三层紫峰缂丝围巾，中间层的金线具有动感，称之为流动的艺术，整个产品具有立体感，轻如羽毛，并获得外观专利与发明专利。2017年，投入大量资金，在保留原有的手工操作技艺上，借用现代智能科技手法，把缂丝最难的技术部分用智能化代替，使操作人员大大缩短了学习时间，只要通过10～15天的基础训练就可以做真品，这一技术的开发将是缂丝行业的一大突破。2018年响应让非遗走进百姓家的号召进行一系列的生活用品的开发，开发生活用品60余种，其中时尚缂丝女包在2017年第四届"紫金奖"文化创意设计大赛荣获金奖，并进行批量生产，年产量在1000件以上。2019年3月，研发了一种新型缂丝机及其相关技术。

Cai Xiaming not only mastered the basic traditional kossu skills, but also mastered the two major techniques — Ming kossu weaving and Ben kossu weaving, which involve material requirements and specifications, drawing manuscripts, color matching, skewing warps, receiving warps, overturning heads, brushing warp surfaces, sticking warp surfaces, shaking threads, weaving and shearing. He created a new artistic expression of works, such as the portrait of *Marilyn Monroe*, which was transformed into cartoons through character images and was woven with kossu fabrics. In weaving, the phase color method of embroidery was applied to the kossu weaving. For example, multi-color kossu waving in oil painting broke the original single-color and two-color phase, and used three-color and four-color combination in rich colors. Through artificial twisting, different effects on kossu weaving were produced. Through research and trial work, she successfully weaved two-sided and three-different skills that no one had mastered since Master Wang Jinshan. In 2018, Cai Xiaming's team developed a three-layer kossu scarf based on the gold silk feather jacket unearthed in Mawendui. The gold thread in the middle layer was dynamic, which was called flowing art. The whole product had three-dimensional sense and was as light as feathers, and had obtained appearance patents and invention patents. In 2017, a large amount of money was invested. By retaining the original manual operation skills, modern intelligent technology was used to replace the most difficult technical part of kossu weaving with intelligence, which greatly shortened the study time of operators. As long as they passed 10 to 15 days of basic training, they could make genuine products. The development of this technology will be a major breakthrough in the kossu industry. In 2018, in response to the call to let intangible cultural heritage enter people's homes to develop a series of daily necessities, and more than 60 varieties of daily necessities were developed. Among them, the fashionable kossu handbag won the gold medal in the 4th Zijin Award Cultural Creative Design Competition in 2017, and they were put into mass-production, with an annual output of more than 1,000. In March, 2019, a novel kossu weaving machine and its related technologies were developed.

3. 成就与荣誉/Achievements and Honors

30多年中，蔡霞明培育了一批批优秀的缂丝技术人才，前后总共80余人，现已分布在各个地区，有现聘请在吴文康缂丝工作室做技术师傅的林英、被评为工艺美术师职称的服装设计师周莹和陆小燕等，其中周莹同学移民德国，在德国开设了自己的绣坊，蔡霞明的多数徒弟已成为缂丝行业中的中流砥柱。

For more than 30 years, Cai Xiaming has cultivated batches of excellent kossu technical talents, with a total of more than 80 people, who are now distributed in various regions, including Lin Ying, currently hired as a technician in Wu Wenkang Kossu Studio, and Zhou Ying and Lu Xiaoyan, fashion designers and arts and crafts artists. Among them, Zhou Ying immigrated to Germany and opened her own embroidery workshop in Germany. Most of the apprentices have become the mainstay of the kossu industry.

蔡霞明于2018年申请专利2项，2017年开始申请版权登记100多项。先后发表《明缂丝与本缂丝》《我对缂丝的探索与思考》《缂丝的历史与现状》《缂丝的制作工艺》等学术论文。缂织的作品获得国内外30余项荣誉。

Cai Xiaming applied for two patents in 2018 and more than 100 patents were applied for copyright registration in 2017. He has published academic papers such as *Ming Kossu and Ben Kossu*, *My Exploration and Reflection on Kossu Fabrics*, *History and Current Situation of Kossu Fabrics* and *Production Technology of Kossu Fabrics*. The kossu works have won more than 30 honors at home and abroad.

2018年，蔡霞明受邀参加"第二十四届世界非遗展"，其作品《玛丽莲·梦露》被主办方永久收藏，围巾《蝶戏牡丹》作为国礼赠予法国总统夫人布丽吉特·马克龙。2019年10月受邀参加由法国文化部、法国国家艺术行业联合会主办的"第二十五届世界非遗展"，作品《兰竹图》被主办方永久收藏。2020年1月11日受中国丝绸博物馆邀请参加"时尚变革中的文化复兴"为主题的"初·新——2019年时尚高峰论坛暨年度时尚回顾展"，作品仿清代故宫宫扇被中国丝绸博物馆永久收藏。

In 2018, he was invited to participate in the "24th World Intangible Cultural Heritage Exhibition" in France. The work *Marilyn Monroe* was permanently received by the organizer, and the scarf *Butterflies and Peony* was presented as a national gift to Brigitte Macron, wife of the French President. In July, October 2019, he was invited to participate in the "25th World Intangible Cultural Heritage Exhibition" sponsored by the French Ministry of Culture and the French National Art Industry Federation, and the work *Orchid and Bamboo* was permanently collected by the organizer. On January 11, 2020, he was invited by China National Silk Museum to participate in the "Beginning, Newness—2019 Fashion Forum and Annual Fashion Retrospective Exhibition" with the theme of "Cultural Revival in Fashion Change". His works imitated palace fan of the Forbidden City in the Qing Dynasty, and were permanently collected by China National Silk Museum.

2016年10月，蔡霞明被苏州市吴中区宣传部等部门评为东吴文化产业重点人才。2017年10月，被苏州市人才工作领导小组办公室等部门评为苏州市民间工艺家。2018年11月，被吴中区文化体育局评为吴中区非物质文化遗产苏州缂丝织造技艺代表性传承人。2019年4月，被吴中区总工会授予"吴中工匠"称号。2019年9月，被中共江苏省委宣传部授予"江苏省紫金文化创意英才"称号。2019年9月，由苏州市吴中区人才工作领导小组办公室授予第二届"东吴杰出匠师"称号。在2020年被授予"江苏省乡土人才'三带'能手"称号。

In October 2016, Cai Xiaming was rated as a key talent in Soochow cultural industry by the Propaganda Department of Wuzhong District of Suzhou City. In October 2017, he was awarded the honorary title of Suzhou Civil Craftsman by Suzhou Municipal Talent Work Leading Group Office and other departments. In November 2018, he was rated as the representative inheritor of Suzhou kossu weaving skills of the intangible cultural heritage of Wuzhong District, by Wuzhong District Culture and Sports Bureau. In April 2019, he was awarded the honorary title of "Wuzhong Artisan" by Wuzhong District Federation of Trade Unions. In September 2019, he was awarded the title of "Jiangsu Zijin Cultural Creative Talent" by the Propaganda Department of Jiangsu Provincial Party Committee. In September 2019, the second title of "Outstanding Craftsman of Soochow" was awarded by the Office of Talent Work Leading Group of Wuzhong District, Suzhou City. In 2020, he was awarded the honorary title of "Three Belts of Local Talents in Jiangsu Province".

二、蔡霞明大师作品/Works of Master Cai Xiaming

2017年，通过研究试作，蔡霞明成功地完成了两面三异的成果，一叶一菩提是佛教中最经典的圣物，因此创作以菩提叶为主题，在一面银地色的菩提叶上缂织了"佛"字，在另一面金地色的菩提叶上缂织了"缘"字，这件作品巧妙地呈现了三件作品《佛》《缘》《佛缘》，如图5-5-1所示。

Through research and trial production in 2017, he successfully completed the two-sided and three-dimensional works. One leaf and one Bodhi are the most classic sacred objects in Buddhism. Therefore, the theme of Bodhi leaves was created, and a Chinese character "Fo" was woven on one side of the Bodhi leaves with silver ground, while a word of "yuan" was woven on the other side of the Bodhi leaves with golden ground. This work skillfully presented three works, *Fo*, *Yuan*, *Fo and Yuan* as shown in Figure 5-5-1.

图5-5-1　两面三异菩提叶
Two-Sided and Three-Dimensional Kossu Bodhi Leaf

◎ 思考题/Questions for Discussion

1．蔡霞明大师掌握的缂丝中的两大技术是什么？/What are the two major kossu techniques mastered by Master Cai Xiaming?

2．蔡霞明大师完成的两面三异的作品是什么？/What are the two-sided and three-dimensional works completed by Master Cai Xiaming?

3．蔡霞明利用光的美学做的作品是什么？/What is the work of Cai Xiaming made by using the aesthetics of light?

4．被中国丝绸博物馆永久收藏的蔡霞明的作品是什么？/What is the work of Cai Xiaming permanently collected by China Silk Museum?

第六节   曹美姐/Cao Meijie

一、曹美姐大师简介/Introduction to Master Cao Meijie

1. 成长历程/Personal Development

曹美姐大师

曹美姐，1957年生于江苏省苏州市。自1980年开始学习缂丝技艺，1985年创办苏州迎春工艺品厂，后更名为苏州工业园区仁和织绣工艺品有限公司，先担任总经理，2016年开始担任艺术总监，同时担任北京曹美姐缂丝文化有限责任公司董事长以及中国缂丝文化公益基金管理委员会主任。目前为高级工艺美术师，在缂丝领域一直不断学习、进修、提升。目前也担任国内高校的客座教授以及项目指导专家。除本人创作外，还授徒、指导技艺等；多年来一直致力于缂丝文化的传承和推广，在技术引领、市场推广、文化推广等方面取得不少成就。

Cao Meijie, born in Suzhou, Jiangsu Province in 1957. Since 1980, she began to learn the kossu skills and in 1985 she founded Suzhou Yingchun Handicraft Factory, which was later renamed Suzhou Industrial Park Renhe Weaving and Embroidery Crafts Co., Ltd. At first, she was served as general manager. Since 2016, she has served as artistic director, and also serves as the chairman of Beijing Cao Meijie Kossu Culture Co., Ltd. and director of the China Kossu Culture Public Welfare Foundation. At present, Cao Meijie is a senior arts and crafts artist who has been constantly learning, training and improving in the field of kossu. She also serves as a visiting professor and project guidance expert at domestic universities. In addition to her own creation, she teaches apprentices and renders skills. Over the years, she has been committed to the inheritance and promotion of kossu culture, and she has made many achievements in technology leadership, marketing promotion, cultural promotion, etc.

2. 对缂丝的研究/Researches on Kossu Fabrics

截至2022年1月，曹美姐本人名字已注册商标，并拥有外观专利9项，实用新型专利12项，发明专利2项，著作权/版权74项；省市新产品鉴定证书2张；带领团队和科研机构合作，攻克了缂丝实用品防水、防油、防污的课题，攻克了缂丝工艺品防霉的课题，技术处于行业领先水平。同时带头起草了《缂丝》行业标准以及参与制定《苏绣》国家标准，并已全国贯标；2019年完成了缂丝学习机的研发，开始与学校、教育机构等合作推广缂丝文化。

As of January 2022, the name of "Cao Meijie" has been registered a trademark, and she has nine appearance patents, twelve utility model patents, two inventions, seventy-four copyrights and two provincial and municipal new product appraisal certificates. Leading the team and cooperating with the research institutions, she overcame the problem of water-proofing, oil-proofing and anti-fouling of kossu practical products, and she also conquered the problem of anti-mildew of kossu handicrafts, which was the technology at the leading level in the industry. At the same time, she took the lead in drafting the "Kossu" industry standards and participated in the formulation of the "Su Embroidery" national standards, which has been implemented nationwide. In 2019, she completed the research and development of the kossu learning machine and began to cooperate with schools and educational institutions to promote the kossu culture.

3. 成就与荣誉/Achievements and Honors

曹美姐将传统工艺与现代时尚相结合，开发了一系列既具艺术性，又具实用性的产品系列；其开发的缂丝礼品系列受到各地政府、外事办公室、侨务办公室以及各个企事业单位的欢迎和好评。

Combining traditional craftsmanship with modern fashion, Cao Meijie has developed a series of products that are both artistic and practical. She explored and developed the kossu gift series, which have been welcomed and praised by local governments, foreign affairs offices, overseas Chinese offices and various enterprises and institutions.

工艺改良，技术上不断创新；在继承传统艺术的基础上，她承古创新，锲而不舍，作品多次在国家级工艺美术大师精品博览会上获奖，并多次受邀到海外进行工艺表演和展示。

She has made technological improvement and continuous innovation in technology. On the basis of inheriting traditional art, she inherits the ancients and innovates with perseverance. Her works have won many awards at boutique fairs of the national arts and crafts masters for many times. Moreover, she has been invited to overseas for craft performances and displays many times.

市场路线调整，构建销售网络；不断打开局面开拓市场，让国人重新认识缂丝，并让缂丝走向国际，使外国友人为中国艺术惊叹。

Through adjusting the market route and building a sales network, Cao Meijie constantly opens up the market, makes the Chinese people re-understand the kossu and makes the kossu become international, so that foreign friends marvel at Chinese art.

培养人才，积极投身于教育事业，与全国各大院校建立合作伙伴关系，为学校教学和学生就业提供帮助。

Moreover, Cao Mejie cultivates talents and actively devotes themselves to education. She establishes partnerships with major universities across the country to provide assistance for school teaching and student employment.

曹美姐获得2007年苏州市"十行百星"、2014年江苏省当代优秀艺术家、2017年江苏省乡土人才"三带"能手等称号。2019年带领企业获评"苏州市未成年人社会实践体验站"，2020年带领企业获评"园区青少年校外教育基地"，建立缂丝培育体系。

Cao Meijie won the 2007 Suzhou Ten Occupations and 100 Stars, the 2014 Jiangsu Contemporary Outstanding Artist, the 2017 Three Belts of Local Talents in Jiangsu Province and so on. In 2019, she led the enterprise to win the title of "Suzhou Minor Social Practice Experience Station". And in 2020 she led the enterprise to obtain the title of "Off-campus Education Base for Youth". Moreover, she also established a system of kossu cultivation.

二、曹美姐大师作品/Works of Master Cao Meijie

迄今，曹美姐大师作品已获得国际金奖1项，国家金奖10项，其他省市各级奖项近40项；缂丝作品被外交部驻外机构供应处选用作为国礼赠送国外元首；缂丝《牡丹》立屏被澳门特别行政区第一、第二任行政长官何厚铧收藏（图5-6-1）；缂丝团扇《墨兰》被著名影星林志玲女士收藏(图5-6-2)。

So far, Cao Meijie's works have won an international gold medal, ten national gold awards, and nearly forty awards at all levels in other provinces and cities. The works have been selected by the Supply Department of Foreign Office of the Ministry of Foreign Affairs as a national gift to foreign heads of state. The standing screen kossu work of *Peony* has been collected by Ho Hau-wah, the Chief Executive of the Macao Special Administrative Region (Figure 5-6-1). The kossu fan *Chinese Cymbidium* has been collected by the famous film star Ms. Lin Zhiling (Figure 5-6-2).

图5-6-1　缂丝《牡丹》台屏
Kossu Ponel *Peony*

图5-6-2　缂丝《墨兰》
Kossu *Chinese Cymbidium*

◎ 思考题/Questions for Discussion

1．曹美姐大师带领团队和科研机构合作攻克了缂丝实用品的哪些课题？/What projects on silk practical products were carried out by Cao Meijie's team and scientific research institutions?

2．曹美姐大师参加过哪些重大项目？/What major projects had Cao Meijie participated in?

第七节　吴文康/Wu Wenkang

一、吴文康大师简介/Introduction to Master Wu Wenkang

吴文康大师

1. 成长历程/Personal Development

吴文康，1960年出生于苏州市高新区东渚镇，自幼受缂丝工艺熏陶，高中毕业后即拜师学艺。四十余年来一直致力于缂丝工艺的研究，在保持缂丝传统特色的基础上，不断创新，让传统的缂丝艺术与现代生活更好地融合在一起，使其兼备了艺术价值与市场价值。作为缂丝传承者，吴文康积极肩负着传承缂丝工艺、弘扬缂丝文化的责任。不仅设立苏州西部民间缂丝织绣厂，而且开设直营店，让顾客与缂丝面对面，近距离感受非遗文化的博大精深；为了使缂丝更好地传承发展，吴文康参加了许多有影响力的艺术博览会，荣获国内外众多奖项，助推了缂丝文化的宣传和弘扬。

Wu Wenkang was born in Dongzhu Town, High-tech Zone, Suzhou City in 1960. He was influenced by the kossu craftsmanship since childhood. After graduating from high school, Wu Wenkang began to study art. For more than 40 years, he has been committed to the research of Kossu craftsmanship. On the basis of maintaining the traditional characteristics of kossu, he has continuously made innovations so that the traditional kossu art and modern life can be better integrated, being of both artistic value and market value. As the inheritor of kossu skills, Wu Wenkang actively shoulders the responsibility of inheriting the kossu technology and promoting the kossu culture. He not only set up Suzhou Western Folk Kossu Factory, but also opened a direct-sale store, allowing customers to face kossu fabrics and experience the profundity of intangible cultural heritage at close range. In order to better inherit and develop kossu, Wu Wenkang participated in many influential art fairs and won various awards, which has promoted the publicity and propaganda of kossu culture.

2007年6月，吴文康受邀参加文化部在北京举办的"非物质文化遗产日"活动，被文化部授予"文化遗产日奖"。同年6月，他所经营的苏州西部民间缂丝织绣厂被江苏省民间文艺家协会授予"江苏省缂丝传承基地"。2009年4月，中国华夏文化遗产基金会与中央电视台拍摄了该厂非物质文化遗产缂丝工艺专题纪录片。2013年，吴文康在香港理工大学设计学院展厅举办了"一寸缂丝一寸金"的展览，此展览吸引了数万人次参观。之后，他收到了澳大利亚、意大利和新西兰等国家设计界的邀请，前往这些国家举办缂丝展览。

In June 2007, Wu Wenkang was invited to participate in the event "Intangible Cultural Heritage Day" held by the Ministry of Culture in Beijing and he was awarded the "Cultural Heritage Day Award" by the Ministry of Culture. In June of the same year, his Suzhou Western Folk Kossu Factory was awarded the "Jiangsu Provincial Kossu Inheritance Base" by the Jiangsu

Folk Artist Association. In April 2009, the China Huaxia Cultural Industry Foundation and CCTV filmed a special documentary on the intangible cultural heritage of the factory. In 2013, Wu Wenkang held an exhibition of "One Inch of Kossu and One Inch of Gold" in the exhibition hall of the School of Design in the Hong Kong Polytechnic University, which attracted tens of thousands of visitors. He then received invitations to hold kossu exhibitions from the design community of countries such as Australia, Italy and New Zealand.

2. 对缂丝的研究/Researches on Kossu Fabrics

三十年的从艺经历，已使吴文康与缂丝艺术融为一体，他复制开发出百余种缂丝名画和织物。为了使缂丝能更好地传承下去，他积极与各大高校联系，美术教授、大学生等先后来工厂里学习缂丝技术。2013年暑假，苏州大学两名学生在吴文康的指导下独立完成了两幅作品。吴文康先生说，他将继续积极与学校合作，让更多的年轻人知道缂丝、了解缂丝、传承缂丝。此外，吴文康与众多设计师合作，不断开发新型缂丝面料，使缂丝与日用品结合，走入日常生活。

Thirty years of artistic experience have made Wu Wenkang fully engaged in the kossu art. He reproduced and developed more than one hundred kinds of kossu paintings and fabrics. In order to better pass down kossu skills, he actively contacted with various universities, and the art professors and students came to the factory to learn Kossu technology. In the summer of 2013, two students of Soochow University independently completed two works under the guidance of Wu Wenkang. Wu Wenkang said that he will continue to actively cooperate with the school to let more young people know kossu, understand kossu and inherit kossu. In addition, Wu Wenkang has cooperated with many designers to continuously develop new kossu fabrics, so that kossu fabrics can be combined with daily necessities and step into our daily life.

3. 成就与荣誉/Achievements and Honors

吴文康十上北京，向中华全国工商业联合会提交材料，希望成立中国缂丝研究会，保护传承缂丝艺术，国家领导十分重视，多次派专家来苏州考察，最终决定成立"中国缂丝研究会"，批文于2013年8月下达，聘请吴文康担任会长。同时，吴文康不忘慈善，与中国扶贫开发协会老区基金会合作，将缂丝产品拿出来拍卖，拍卖所得全部作为善款。

In order to protect and inherit kossu art, Wu Wenkang went to Beijing for ten times and submitted materials to the All-China Federation of Industry and Commerce, hoping to establish the China Kossu Research Association to protect, inherit and develop the art of Kossu. The national leaders attached great importance to it and sent experts to Suzhou for many times. Finally, the approval document was issued in August 2013 to establish the "China Kossu Research Association". And Wu Wenkang was served as the president. At the same time, Wu Wenkang did not forget charity and cooperated with the Old District Foundation of the China Association for Poverty Alleviation and Development to auction the kossu fabrics and get all the money for donations.

吴文康通过多年的努力，对缂丝工艺做出杰出的贡献，被联合国教科文组织授予"中国民间工艺美术家"的称号，他的业绩被载入《中国民间文艺家大辞典》《共和国建设者档案》《苏州民间手工艺术》等。

Through years of efforts, Wu Wenkang has made outstanding contributions to kossu craftsmanship and has been awarded the title of "Chinese Folk Arts and Crafts Artist" by UNESCO. His achievements have been included in *the Dictionary of Chinese Folk Artists*, *the Archives of the Builders of the Republic* and *the Suzhou Folk Handicraft Art*.

吴文康为研究员级高级工艺美术师、江苏省工艺美术名人、江苏省乡土人才"三带"能手、苏州缂丝织造技艺代表性传承人、苏州工艺美术大师、中国民间文艺家协会会员、苏州市缂丝协会副会长、苏州市民间文艺家协会副秘书长、苏州市民间文艺家协会织绣专业委员会副会长以及苏州市工艺美术学会会员。

Wu Wenkang is the researcher-level senior arts and crafts artist, a celebrity of Arts and Crafts in Jiangsu Province, a celebrity of Three Belts of Local Talents in Jiangsu Province, a representative inheritor of Suzhou kossu skills, a master of Arts and Crafts in Suzhou, a member of the Chinese Folk Writers Association, a vice president of the Suzhou Kossu Association, a deputy secretary-general of the Suzhou Civic Writers Association, a vice president of the Weaving and Embroidery Professional Committee of the Suzhou Civic Writers Association and a member of the Suzhou Arts and Crafts Society.

二、吴文康大师作品/Works of Master Wu Wenkang

吴文康大师的作品如图5-7-1至图5-7-3所示。在多年的从艺经历中，吴文康越发仰慕传统缂丝艺术，自发复原了多件明清时皇帝的龙袍，龙袍上团龙鳞片纹路清晰，双目炯炯有神，从古典的缂丝精品中，充分拾得历史的底蕴与精髓。

The works of Master Wu Wenkang are shown in Figures 5-7-1 to 5-7-3. With the years of artistic experience, Wu Wenkang admired the traditional kossu art increasingly. He spontaneously restored a number of dragon robes of the Ming and Qing emperors with clear pattern of dragon scales and bright eyes

图5-7-1 缂丝龙袍
Kossu Dragon Robe

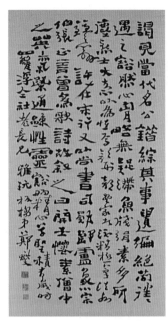

图 5-7-2 缂丝《韩熙载夜宴图》
Kossu *Night Revels of Han Xizai*

图 5-7-3 缂丝郑板桥书法
Kossu Zheng Banqiao's Calligraphy

on the dragon robes. He found the historical heritage and essence from the classical kossu fabrics.

◎ 思考题/Questions for Discussion

1. 作为缂丝传承者，吴文康是如何积极肩负着传承缂丝工艺、弘扬缂丝文化的责任的？/As an inheritor of kossu art, how does Wu Wenkang actively shoulder the responsibility of inheriting and promoting the silk kossu culture?

2. 吴文康的从艺经历是怎么样的？/What is Wu Wenkang's artistic experience?

3. 吴文康所经营的苏州西部民间缂丝厂被江苏省民间文艺家协会授予什么的称号？/What title was awarded to Wu Wenkang's Suzhou Western Folk Kossu Factory by the Jiangsu Folk Writers Association?

第八节　陈文 /Chen Wen

一、陈文大师简介/Introduction to Master Chen Wen

1. 成长历程/Personal Development

陈文大师

陈文，1969 年出生，苏州人。1989 年起从事缂丝的图案设计，并深入学习缂丝技艺和全程监制，从而执着于缂丝的传承和发展。2001 年创办

"祯彩堂"，带领旗下50位手艺师傅，从近代缂丝起落中走出，打造全新的缂丝品牌——祯彩堂。多角度诠释缂丝的艺术之美和实用之道，致力于缂丝的良性传承。作品曾在省级、国家级评选中获得众多奖项。

Chen Wen, a native of Suzhou, was born in 1969. Since 1989, she has been engaged in the design of the kossu patterns, in-depth learning of kossu techniques and supervision of the whole process, so as to be persistent in the inheritance and development of Kossu. In 2001, she founded "Zhen CaiHall" and led its 50 craftsmen to emerge from the ups and downs of modern Kossu to create a brand-new Kossu brand—Zhen Cai Hall, which presents an interpretation of the artistic beauty and practicality of Kossu from multiple angles and commits to the virtuous inheritance of kossu skills. Her works have won many awards at the provincial and national levels.

2. 对缂丝的研究/Researches on Kossu Fabrics

由于制作精湛，陈文团队的缂丝作品受到广大缂丝爱好者的欢迎，同时陈文也成功申请了缂丝手包墨彩花卉等外观专利6项，实用新型专利1项（一种便于学习的轻便可折叠缂丝织机），发表论文1篇。

Due to the exquisite production, the kossu fabrics of Chen Wen's team have been welcomed by the kossu lovers. Chen Wen has also successfully applied for six appearance patents, such as kossu handbag with ink-colored flowers, one patent of utility model (a lightweight and foldable kossu loom that is easy to learn) and one paper published.

3. 成就与荣誉/Achievements and Honors

陈文善于将色彩、图案及缂丝技艺很好地进行融合，其设计的作品典雅、秀美、精致，体现缂丝之美。

Chen Wen is good at blending colors, patterns and kossu techniques. Her designed works are elegant, beautiful and exquisite, reflecting the beauty of kossu fabrics.

同时，陈文致力于缂丝在中小学生中的传承，她与苏州市实验小学校等学校合作，通过社团等形式，教授缂丝技艺，将学校与社会良好地融合在了一起，理论与实践结合，彰显了世界非物质文化遗产的魅力，也为传统手工艺的现代传承提供了新渠道。

At the same time, Chen Wen is committed to the inheritance of kossu art among primary and secondary school students. Through the form of clubs and other forms, she cooperates with Suzhou Experimental Primary School and other schools to teach kossu skills, integrating school education with the social culture. Combining theory with practice, it highlights the charm of the world's intangible cultural heritage and provides a new channel for the modern inheritance of traditional handicrafts.

陈文现为高级工艺美术师、苏州缂丝织造技艺市级代表性传承人、苏州工艺美术学会缂丝专业委员会秘书长，获苏州市工艺美术大师称号。

At present, she is a senior arts and crafts artist, a municipal kossu inheritor of intangible cultural heritage, the secretary general of the Professional Committee of Kossu of Suzhou Arts and Crafts Society and the Master of Arts and Crafts in Suzhou.

二、陈文大师作品/Works of Master Chen Wen

从最近陈文大师的"方寸"祯彩堂缂丝艺术展中可以看出，其作品主要可分为"千年遗脉""文房雅趣""艺家生活"三个方面。"千年遗脉"主要为古代缂丝作品的复制品，如唐代的缂丝带、北宋的《紫鸾鹊谱》、南宋的《山茶蛱蝶图》等典型作品。

It can be observed from Chen Wen's recent "An Inch" Zhen Cai Hall Kossu Art Exhibition, her works can be mainly divided into "Millenary Relics" "Elegance in the Study and "Artist Life". The "Millenary Relics" are mainly reproductions of ancient kossu fabrics, such as the kossu ribbon in the Tang Dynasty, *Phoenixes and Magpie with Purple Background* of the Northern Song Dynasty, *Camellia and Butterflies* of the Southern Song Dynasty and other typical works of this era.

"文房雅趣"主要是文房中的缂丝用品，体现出一方文雅天地，展示的作品构图严谨、色泽和谐、形象生动、栩栩如生。从包裹珍贵书画作品的装裱包首件，到直接摹织书画，缂丝作品始终与文房陈设形影相随。祯彩堂琢磨文人闲情逸趣，极致表现画意，缂制的书画作品突破了前代的工笔题材，内容更为宽泛，拓展了缂丝所能呈现的艺术效果。包含书法与写意画的水墨类作品、花卉类作品是祯彩堂缂丝书画系列的两大特色。经过匠人的缂织，原本二维的笔墨色彩，增加了织物的纤维质感，运丝传神。主要作品为《岁朝丽景图》《竹石图》《六柿图》《太湖人家》《水墨画册页》《宫扇荷花》等。如图5-8-1所示，《太湖人家》作品历时两年才制作完成。用缂丝技艺表现泼墨山水的难度极大，推敲练磨、反复试色的时间很长。作品采用拼线、合花线、戗色技法，运用灵活，收放自如，天衣无缝，较完整地保留了大千山水的笔墨神韵，是难得的佳作。

"Elegance in the Study" is the major kossu product in the study, reflecting the elegance of the world. The works displayed are rigorous in composition, harmonious in color, and vivid in image. From the first piece of the mounting bag that wraps the precious calligraphy and painting works, to the direct copy of the calligraphy and painting, the kossu works have always been matched with the furnishings of the study. Framed bag wrapped in precious calligraphy and painting works to the direct imitation of calligraphy and painting, the kossu products are always similar to the furnishings of the study room. Zhen Cai Hall pondered leisure of the literati and reveals the essence of paintings. The production of Zhen Cai Hall and painting works broke through the themes of the previous generation of fine brushwork, with broader contents and the expanding artistic effect. Ink works and floral works are the two major features of the Zhen Cai

图5-8-1　缂丝《太湖人家》
Kossu *The Family by Taihu Lake*

Hall kossu Calligraphy and Painting Series, including calligraphy and freehand paintings. After the craftsman's kossu weaving, the original two-dimensional pen and ink color increased the fiber texture of the fabric. From a distance, it is a painting. From the close touch, it is for appreciation of taste, and for appreciation of the magic of the weaving shuttle. His main works are *Beautiful Scenery of the Dynasty, Bamboo and Stones, Six Persimmons, The Family by Taihu Lake, Ink Painting Album, Palace Fan and Lotus* and so on. As shown in Figure 5-8-1, *The Family by Taihu Lake* took two years to complete. It is extremely difficult to display the splashed ink landscape with kossu techniques. The time for repeated testing of colors is very long. The works use stitching, combining threads and coloring techniques, which are flexible in use, freely retractable, and skillful. The work is a rare masterpiece.

　　曾经的皇家御制为缂丝注明了高级定制的标签，而回归当下生活才是缂丝良性传承必不可少的先决条件。高雅的家居艺术饰品，成为祯彩堂缂丝用品设计的主流方向。于是，祯彩堂与艺术圈人士、博物馆文创设计师和其他门类设计师跨界合作，从传统文化基因到艺术家原创素材，提取最适合缂丝表达的内容进行二次创作。同时细心体味生活，寻找切入点，将缂丝融入生活的方方面面。缂丝用品的发展空间在祯彩堂被拓展至无限的可能。"艺家生活"作品主要包括钱包、手包、手袋、茶席、服饰、团扇、屏风等一系列富有生活美学和时代感的缂丝日用产品。

　　Once as the royal ordered products, kossu is regarded as the label of advanced customization fabrics. Returning to the present life is an indispensable prerequisite for the virtuous inheritance of kossu fabrics. Elegant home decorations have become the mainstream direction of Zhen Cai Hall kossu products design. Therefore, Zhen Cai Hall cooperated with people in the art circle, museum cultural and creative design and other designers. From traditional cultural genes to artists' original materials, Zhen Cai Hall extracted the most suitable content for secondary creation. At the same time in order to integrate kossu into all aspects of people's lives, they looked for entry points and carefully taste life,which the development space of kossu products has been expanded to unlimited possibilities in Zhen Cai Hall. Works of "Artist Life" mainly include wallets, handbags, handbags, tea parties, clothing, round fans, screens and a series of daily kossu products rich in life aesthetics and sense of the times.

图5-8-2所示的水墨屏风这一系列作品取材于水墨画家的花卉小品，构图充分体现画意空灵，看似随性，其实笔笔有气韵流动。缂丝写意具有极高挑战性，通过金属线和丝线混合，既增加质地的牢固度，又营造文雅书卷气氛，呼应当下生活需求，实现生活艺术化的境界愿景。

The series of works of the ink screen is based on the floral sketches of the ink painter shown in Figure 5-8-2. The composition fully reflects the ethereal meaning of the painting, which seems to be casual. But in fact, the brush strokes are flowing. Through the mixing of metal threads and silk threads, the kossu expressing freehand brushwork is extremely challenging, which not only increases the firmness of the texture, but also creates a gentle and elegant bookish atmosphere, echoing the current needs of life and realizing the vision of an artistic life.

图5-8-2　水墨屏风
Screens with Ink Painting

此外，还有一系列缂丝作品，如图5-8-3至图5-8-8所示。如2014年远山系列团扇；2015年远山系列票夹、远山系列茶席；2016年远山系列七巧屏风；2017年远山装置"渡境"。艺术装置"渡境"入选参加在苏州金鸡湖美术馆举办的"机杼——当代艺术展"并获得"苏艺杯"金奖，远山缂丝团扇入选参加在杭州中国扇博物馆举办的"明月入怀·中国团扇文化印象展"；2018年中国远山屏风参加在中国国家博物馆举办的"2018年中国当代工艺美术双年展"，并被中国工艺美术馆收藏；远山系列获得"2018年度江苏省优秀版权二等奖"；2018年远山首饰参加"淘宝造物节"；2019年，与飞利浦共同打造远山系列非遗定制版插座。

In addition, there is a series of works, as shown in Figures 5-8-3 to 5-8-8. For example, there are the 2014 round fan of Distant Mountain series, the 2015 ticket holder of Distant Mountain

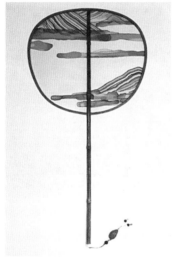

图5-8-3　远山系列团扇
Round Fan of Distant Mountain Series

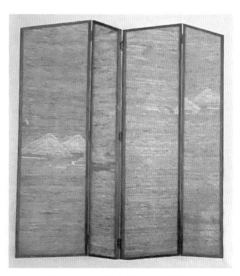

图5-8-4　远山系列七巧屏风
Tangram Screen of Distant Mountain Series

图5-8-5　远山系列茶席
Tea Party of Distant Mountain Series

图5-8-6　远山系列首饰
Jewelry of Distant Mountain Series

图 5-8-7　远山系列非遗定制版插座
Socket of Intargible Customized Version of Distant
Mountain Series

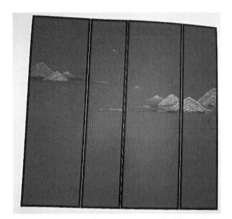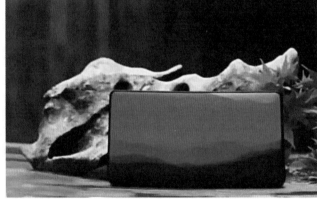

图 5-8-8　远山屏风（别样红）及坤包
Distant Mountain Screen (Different Red) and Handbag

series, tea party of the Distant Mountain series, and the 2016 tangram screens of Distant Mountain series. The Distant Mountain installation "Crossing the Border" was created in 2017. The art installation "Crossing the Border" was selected to participate in the "Contemporary Art Exhibition of Loom" held at Suzhou Jinji Art Lake Museum. The art installation "Crossing the Border" won the gold award of "Su Yi Cup". The round kossu fan of Distant Mountain series was selected to participate in the "Breezing in the Moonlingt Impressions of Chinese Grcular Fans" held at the China Fan Museum in Hangzhou. In 2018, Distant Mountain Screen participated in the "2018 Contemporary Arts and Crafts Biennale" held at the National Museum of China and was collected by the National Art Museum of China. The Distant Mountain series won the "Second Prize of Excellent Copyright in Jiangsu Province in 2018". In 2018, the Distant Mountain Jewelry participated in the "Taobao Maker Festival". In 2019, Distant Mountain series, authorized Philips and jointly created the Distant Mountain Series intangible cultural heritage customized version socket.

◎ **思考题/Questions for Discussion**

1．"文房雅趣"中的作品有哪些？/What are the works of "Elegant in the Study"？

2．"艺家生活"的主要作品有哪些？/What are the main works of "Artist Life"？

3．陈文大师获得"苏艺杯"金奖的是什么作品？/What is the work of Chen Wen that won the Gold Award in "Su Yi Cup"？

4．陈文大师获得"2018年度江苏省优秀版权二等奖"的是什么系列？/Which series of Chen Wen won the Second Prize of Excellent Copyright in Jiangsu Province in 2018?

参考文献／References

［1］朴文英. 中华锦绣丛书：缂丝［M］. 苏州：苏州大学出版社，2010.

［2］濮军一. 中国工艺美术大师王金山缂丝［M］. 南京：江苏美术出版社，2013.

［3］马惠娟，胡金楠. 吴中绝技中国缂丝［M］. 扬州：广陵书社，2008.

［4］赵丰，屈志仁. 中国丝绸艺术［M］. 北京：外文出版社，2012.

［5］赵丰. 敦煌丝绸艺术全集：英藏卷［M］. 上海：东华大学出版社，2007.

［6］张俐敏，徐梦宇，张子阳，等. 基于克孜尔石窟壁画灵感的现代缂毛装饰织物设计与开发［J］. 毛纺科技，2019，49（1）：22-24.

［7］新疆维吾尔自治区博物馆. 新疆吐鲁番阿斯塔那北区墓葬发掘简报［J］. 文物，1960(6)：13-21.

［8］邵晨霞. 缂丝服饰品的传承与发展探析［J］. 丝绸，2010（10）：50.

［9］夏荷秀，赵丰. 达茂旗大苏吉乡明水墓地出土的丝织品［J］. 内蒙古文物考古，1992（1-2）：113-120.

［10］朴文英. 缂丝的起源与传播［J］. 辽宁省博物馆馆刊（第3辑），2008（3）：483-484.

［11］沈国庆，黄俐君. 浅析辽代水波地荷花摩羯纹绵帽的缂丝工艺［J］. 丝绸，2003（3）：46-47.

［12］李超德. 缂丝传统手工艺传承与创新的几点体会：以书画缂丝团扇设计与制作为例［J］. 装饰，2014，253（5）：38-43.

［13］朴文英. 两件缂丝牡丹赏析［J］. 文物，2001（1）：29-30.

［14］马玥婷. 评述宋代缂丝的制作技术和艺术价值［J］. 今日南国，2009，148（2）：145.

［15］茅惠伟. 元代服用缂丝［J］. 丝绸，2006（7）：49-51.

［16］廖军，许星. 论缂丝的艺术特色及其开发［J］. 丝绸，2005（11）：47-48.

［17］杨烨. 织中之圣：中国缂丝的传统技艺传承［J］. 中华文化论坛，2013（3）：141-144.

［18］曾梦娜. 传统缂丝工艺在现代高级定制女装中的运用研究［J］. 西部皮革，2019（2）：48.

［19］梁瑞丽. 苏州丝绸"文创"路新探索［J］. 中国纺织，2016（5）：90.

［20］李建亮. 论苏州缂丝的艺术特色［D］.苏州：苏州大学，2007.

［21］张楠. 缂丝在现代服装设计中的应用研究［D］. 上海：上海工程技术大学，2020.

［22］王喆. 苏州缂丝的多元化视觉传播研究［D］. 无锡：江南大学，2020.

［23］郜莉. 苏州缂丝的现代传承与发展研究［D］. 苏州：苏州大学，2014.

［24］连晓君. 苏州"祯彩堂"缂丝艺术的发扬性保护与管理之研究［D］. 南京：南京航空航天大学，2017.

［25］崔文博. 宋代绘画对传统缂丝工艺的影响［D］. 苏州：苏州大学，2009.